BIGFOOT INFLUENCERS

CANDID CONVERSATIONS WITH RESEARCHERS, SCIENTISTS, AND INVESTIGATORS

TIM HALLORAN

HANGAR 1 PUBLISHING

For my wife Dana, thank you for leading me on this journey and supporting me every step of the way. You inspire me to not let anything stand in my way of accomplishing my goals.

For John Green and René Dahinden, you laid the foundation for today's researchers and your contributions shaped the history of the subject.

Copyright © 2022 by Tim Halloran

All rights reserved.

No part of this book may be reproduced in any form or by any electronic or mechanical means, including information storage and retrieval systems, without written permission from the author, except for the use of brief quotations in a book review.

CONTENTS

FOREWORD BY CARLOS JIMENEZ

So much proof exists for the existence of Bigfoot that the preponderance of the evidence is an embarrassment of riches; surprisingly, though, most people are not aware of this verifiable fact.

I want the truth of my own and countless others' experiences with Bigfoot to be known and understood as proof of their existence--proof that when coupled with the forensic and other physical evidence is determined to be statistically significant data to which Occam's razor can only lead the reader to one conclusion: Bigfoot do exist in our natural world.

My goal in getting involved in this undertaking has always been for the purpose of full disclosure by all stakeholders involved in the Bigfoot "phenomenon"—from those who have encountered them and know that they exist to the opposite end of the spectrum of those who are involved in misinformation or any other subterfuge to keep the public in the dark about a worldwide population of unknown hominids—whether intentionally or unintentionally. Author Tim Halloran shares this goal in this seminal work on some of the key contributors in the Bigfoot world who know that the beings exist and are working tirelessly to gather more definitive evidence for the rest of the public.

Not only am I one of the stakeholders written about in this book, but I also have networked with many of the others whose incredible stories you will read about, and I have also known several of the individuals who paved the way for us and future investigators on the authenticity of Bigfoot. Some have referred to my former professor Dr. Grover Krantz, journalist John Green, outdoorsman and author René Dahinden, and big-game hunter Peter Byrne as the Four Horsemen of serious Bigfoot investigations. I had the pleasure of meeting three of these men while they were still alive and can share a little about what I learned from them.

I met René Dahinden and John Green in 1996 in Harrison Hot Springs, British Columbia where Dr. Krantz had sent me to present my preliminary Anthropology Master's Degree thesis on Gigantopithecus as a possible fossil ancestor of Bigfoot. Although he had many positive contributions to our knowledge of Bigfoot through documentation and evidence he had collected over the years, my first experience with Dahinden was hearing and seeing his contempt for some of the presenters while I was in the audience: He was very vocal and animated in this regard toward so-called academics. Ironically, one of the funniest things that happened during my time at the Sasquatch Days Festival was when the event organizers jokingly bestowed an honorary PhD on Dahinden, as he was cantankerously callous towards those with formal education like my mentor Dr. Krantz.

I had a lovely meal with John and June Green at their home while I was there. We talked at length about his years of researching, investigating, and writing about Bigfoot. I interviewed him about his database (which he offered me on an old 5 ¼-inch floppy disk) and questioned him at length about several of the tales in his publications. One tale was the story of Albert Ostman which he shared in "On the Track of Sasquatch." I was not convinced that this was a true account at the time, as I had not come across any corroborating anecdotes like it. John unequivocally presented the details of his interview and the veracity of Ostman's account, and I left his home that night less doubtful about the details of that story than ever and believing that he was a man of integrity when it came to his investigative work.

I first met Dr. Krantz during my senior year at Pomona College where I was doing my graduation project in Anthropology on Bigfoot under Dr. James McKenna. For my project, I flew up to Pullman, Washington to interview him about his book, "Big Footprints: A Scientific Inquiry into the Reality of Sasquatch." I still have that interview on a couple of old video cartridges somewhere. After interviewing Dr. Krantz and finding him to be very knowledgeable academically about Bigfoot and other aspects of Physical Anthropology, I decided to apply to graduate school at Washington State University (Wazzu) to pursue my advanced degree in Physical Anthropology and conduct scientific research and investigations on the topic of Bigfoot. I was accepted and approved to write my master's thesis on Bigfoot with an emphasis on their possible fossil origins.

During my years at Wazzu, I interviewed eyewitnesses, learned to make molds and casts of bones and footprints, and had my second Bigfoot encounter in the Blue Mountains. Every day before class, I would pepper Dr. Krantz with questions and ideas that I had for different scientific approaches to proving the existence of Bigfoot. I was a young idealist with ambitious ideas that would have been feasible with more funding, but Krantz always listened to me and would often agree or disagree simply by tilting his head one way or the other and raising a brow or two. A funny, albeit macabre, story I can share was from one time when I was in his Osteology lab for class. He said, "Today, you will get to meet my kids." My classmates and I looked at each other in disbelief, as we didn't know that he had any kids. That's when he emptied a box of assorted children's bones donated to science from a mass grave in Chiapas Mexico. Krantz had a very dry sense of humor—when he had a sense of humor because most of the time he was very serious about his teaching.

The fourth "horseman" was one I never met in person, Peter Byrne, but luckily, he is the only horseman who is still alive and was able to be interviewed by the author for this book, about which I will now give you my wholehearted recommendation.

Tim started from the outside looking in, as he had no interest in Bigfoot but was taken to a Bigfoot conference in Ohio by his wife who had always been interested in the phenomenon. There he was impressed by the academic presentations of Jeff Meldrum and Cliff Barackman; he appreciated the true nature of the science and preponderance of evidence for Bigfoot's existence. He learned that a great deal of scientific processes, tactical approaches, and time had been invested in the quest for the truth about Bigfoot. Filled with enthusiasm, the author wondered how anyone would know who was doing what in the way of Bigfoot investigations. Everyone seemed so insulated from one another. Tim was not a field researcher, nor did he intend to become one. He decided to leap into the Bigfoot phenomenon and possibly, unintentionally, becoming an influencer himself by interviewing people and creating this book. Tim knows the mountain of evidence validating that there is something there to investigate.

The author is creating a sort of reference guide of current Bigfoot science—the amazing people actively conducting serious research or affecting the subject and the factors motivating them. It highlights and gives an overview of the individuals conducting and influencing the science of Bigfoot. With the influencers, Tim dives into their theories, experiences, and research, giving readers a glimpse into their historical impact on the subject and what their current projects are. Tim did his due diligence in choosing the interviewees for the book. It provides an unbiased summary of the leaders in the field and their enormous and strategic efforts. The author presents the value of knowing diverse, impactful research approaches and findings and his admiration for the researchers he has met. The book provides distinct viewpoints from people who have witnessed Bigfoot in their life. The author wants to raise awareness and open channels of communication—creating dialogues between all those vested in the truth. This book answers the need to start to get to know who is doing what, where, and how currently in the Bigfoot world—to get a better understanding of what each researcher is doing with their own body of work.

Tim realized that many researchers are disconnected from each other, often working in small groups within their local region. We need the various, serious Bigfoot investigators to work together, sharing data, processes, and tips with one another to advance our shared goal of full disclosure to the public of still yet an unproven hominid species. Influencers are more open to helping others in a unified effort to shape our knowledge. This is more productive for the cause/ research collaboration.

The first book will be a general overview of several key investigators in North America. Features in the book include including QR codes that will allow readers to use their mobile devices to hear audio, see photos, and hear and see video samples from the researchers. The chapters are meant to provide a snapshot of the influencers with information on how to connect with them. Tim wants to create a network so Bigfoot witnesses and those interested in serious Bigfoot inquiry can collaborate. With a new wave of curious younger people that are open-minded about unexplained phenomena like Bigfoot, UFOs, paranormal, etc., the author's goal is to nurture collaboration in the Bigfoot field not only for the sake of the investigators but for the benefit of these new interested parties. Subsequently, the book will have resource recommendations for those new to the Bigfoot phenomenon.

Bigfoot researchers do what they do for the knowledge and love of the discovery—not for fame or fanfare but for validating their stories and personal encounters. Tim has become a fan of these researchers and their serious commitment to the study of Bigfoot. The challenge is in how to not leave someone out—to be as inclusive as possible without being exhaustive. This will allow for adding additional researchers in future volumes. Whether you are an enthusiast or a researcher, The Bigfoot Influencers shares fascinating experiences and facts about the influencers in North America and the impact they are having. His future volumes will likely have regional focuses throughout North America.

Carlos Jimenez, B.A., M.A. Anthropology

Thechroniclesofcarlos.com

cj@chroniclesofcarlos.com

Gill, George W. and Jimenez, Carlos J., 'Paleoamerican skeletal features surviving into the Late Plains Archaic', in Lepper, Bradley T. and Bonnichsen, Robson (eds.), New Perspectives on the First Americans, College Station: Texas A & M University Press, 2004, pp. 137–141, Google Scholar, 138.

INTRODUCTION

The historical accounts of large, hairy ape/man-like creatures date back centuries and are shared by multiple cultures throughout North America. Reports are documented from Florida to Alaska. We commonly refer to them as Bigfoot, Sasquatches, and Wood Apes, amongst other regional labels for them. These "Hairy Men" as described, have also been documented in more than 90 Native American and First Nations cultures, each with their own name and sharing similar descriptions. These accounts are documented through traditions, folklore, art, and rock pictographs dating back 1000 years. Since these stories exist in such a vast number of the native cultures, all without the benefit of communicating through social media, newspapers, telephones, and such like we have today, it makes a compelling case that they may have existed throughout recent history. For this reason, most researchers involved in the subject of Bigfoot, feel these historical indigenous accounts are amongst the strongest forms of evidence supporting their existence.

In modern times and in pop culture we see the terms Bigfoot and Sasquatch associated with advertising, television, and movies. So where did these names originate? The term "Bigfoot" was created in the United States during the 1950s. Daniel Perez, who is a historian on the subject, shares how the term "Bigfoot" was coined:

"Most researchers who have written about how the term 'Bigfoot' was coined usually come close although they miss the mark. You have to go back to the primary source, Andrew Martin Genzoli, (1914 - 1984). In the Humboldt Times (Eureka, California) newspaper for October 1958 the journalist would write, 'The big foot, as cast, was 16 inches in length, with the wide seven-inch width. [Jerry] Crew said the men [the road builders] refer to the creature as 'Big Foot.'

'Big Foot' was later compounded (Bigfoot) and changed to simply 'Bigfoot' and has been part of the lexicon ever since 1958. So, by his own admission, Andrew Genzoli DID NOT coin the term 'Bigfoot,' the real attribution would have to be credited to the road builders who were pushing a road into the Bluff Creek area of northern California for logging.

Another reporter by the name of John Willard Chamberlin who died at age 71 on March 21, 1955, may have been the first to use the actual term "Big foot" in a newspaper article in 1955. Although John Chamberlin may have beat everyone to the punch, the use of that term then, in relation to some tracks that were found, did not catch on like it did in 1958."

"Sasquatch" earned its name in the record books decades prior. Canadian researcher and historian, Thomas Steenburg shared how the term was coined:

"In the West coast of Canada and the United States, there were multiple names for them, both First Nation and Caucasian. In the late 1920s, a man by the name of J.W. Burns took a job with the Sts'ailes First Nation people, he was an agent for the *Crown.* The people shared their mythology and folklore and started telling him about these things in the woods they called *Sas'qets.* J.W. Burns also authored articles that would appear in Canadian magazines. He wrote one that appeared in *Maclean's* on April 1st, 1929. It was titled

Introducing B.C.'s Hairy Giants. In that article, he misspelled *Sas'qets* and called them "Sasquatch". It's been known as Sasquatch in Canada ever since".

Publicized Bigfoot expeditions in North America go back prior to the 1920s and older accounts can be found in media archives. In the 1950s though, the search for the elusive Bigfoot seemed to elevate and today thousands of "citizen scientists" continue to scour the woods in attempts to solve the mystery. Some of the pioneers of that time who paved the way for today's researchers include Peter Byrne, Rene Dahinden, John Green, Grover Krantz, John Bindernagel, Bob Titmus, Vladimir Markotic, Jerry Crew, Roger Patterson, Bob Gimlin, and Wes Summerlin. In 1967, the Patterson-Gimlin film was shot in Northern California and is the most famous video related to the subject of Bigfoot. Furthermore, most Bigfoot researchers and enthusiasts feel it is the most compelling evidence in existence today.

The goal of this book, *The Bigfoot Influencers*, is to provide a summation of some of today's scientists, influencers, and researchers who are studying this phenomenon. How did I define *Influencer* as it relates to this book? An influencer is an individual or a group that has made a historical impact, done notable work, or is affecting how others view the subject of Bigfoot. There are many researchers who are not with us today (some noted above), and the purpose of this book is to deliver a glimpse of the current individuals influencing this subject, their backgrounds, personal interests, and theories on the topic. In addition to exploring the years of focused attention these individuals have dedicated to the subject, all have magnanimously shared their theories and practices to support the goals of this book. Though many feel the "Bigfoot Community" lacks collaboration, based on my experience since taking on this project, I refreshingly found that this was not the case. Similar to you and I, the individuals in this book have families, interests, and hobbies outside of Bigfoot. I hope you learn something new about them, get a few laughs, and maybe get introduced to some people that you may not have previously followed.

There are hundreds of worthy researchers that I could incorporate into this book, and I plan to include additional individuals and groups in the upcoming volumes. I encourage you to dive in with an open mind and enjoy the forthcoming chapters.

A unique feature we have added is the use of *QR Codes*. Readers will find these located throughout the book. Each *QR Code* will allow you, the reader, the ability to scan each one using the camera feature on your mobile device. From there, your mobile device will immediately link to additional fascinating videos, audio, images, or information pertaining to the given section of the book you are delving into.

Open Camera on mobile device
and point at QR Code

Les Stroud Introduction

CLIFF BARACKMAN

Cliff Barackman may be best known to the public for his role in *Finding Bigfoot*, which may be the most successful TV series on the subject, although his impact on Bigfoot research has been far more expansive. He has been involved in Bigfoot research for almost 30 years. He and his wife Melissa opened and run the North American Bigfoot Center, in Boring, Oregon. Cliff has participated in numerous field projects around the world, has appeared on multiple TV documentaries and co-hosts the popular podcast *Bigfoot and Beyond* with James 'Bobo" Fay. Cliff's primary research is focused on studying foot and handprints. I would suggest visiting his website to review his footprint database, Cliff provides historical images and detailed records of what is known about them.

I met Cliff for the first time a few years ago at the Ohio Bigfoot Conference, he was pleasant, welcoming and took the time to chat with all who approached him. Although he claims to be an introvert, Cliff is very witty, an excellent communicator and has a natural sense of humor. Cliff is passionate about ensuring that his Bigfoot research and advocacy stay focused only on the animal. My interview with Cliff was enjoyable, one of my favorites, I appreciated his openness and candidness as we dove into the topic of Bigfoot. We also found ourselves expanding the conversation to philosophy, gardening, and music, which was really cool.

Cliff half-jokingly states that he has gotten grumpy over the years which may weigh in on his opinions regarding people, society, and the *Bigfoot Community*. I found him to be generous, loyal and someone who cherishes relationships with his friends. Historically, Cliff was part of a team that brought in a new generation of Bigfoot enthusiasts and researchers whom many of them will continue to advance awareness on the subject into the future.

What are your hobbies? I have a degree in jazz guitar and enjoy playing as much as possible, I try to get in a few hours a week of playing. It is good for my head, it's cerebral music. I like to read and love gardening; I have a 23-acre property in the woods so I enjoy spending time outdoors. I just planted garlic for the winter months, these are my aspirations in life [laughing]. There is something deeply satisfying about gardening and that kind of thing. In some of my more frustrating public moments, I have often thought that my legacy may be a strong relationship in a good garden and that would be satisfying enough. There is a little bit of a Sam Gamgee in all of us. I also cook, I am just kind of a roundabout guy, I play tabletop roll playing games, I have all sorts of whacky interests [laughing].

How did you get into music? Well, I was young, 12 or 13 years old, and it seemed everyone else on my street enjoyed playing football. My friends didn't like me to play because I didn't care about winning, I was just out there goofing around and having fun. Then at some point, I picked up my dad's guitar which was laying around the house, it was an old classic one. I was not really a social person to begin with, so instead of playing sports with my friends, I would sit in my room and learn chords and that started the journey.

The enjoyment of playing music is that you are interacting with a group of musicians in *the moment* for the purpose of *that moment*, if that makes sense. There is something fantastic, life-affirming and spiritual about that. Occasionally if everything is lined up right, magic happens and you cannot explain it.

Who are your favorite musicians or bands? I love old country; Patsy Kline, Johnny Cash, etc... I grew up on the 80s synth-pop music: Depeche Mode, etc.... A band that really bent my ear early was Oingo Boingo because they are just so unusual, and produce heavy, smart music. Another fantastic band is the Sparks Brothers. For the most part, they flew under the radar, although a documentary film was recently made about them. They are just fantastic, sarcastic, weird, and quirky. I am also a huge fan of Steely Dan, you can sometimes hear their music in elevators and supermarkets, and they are singing about smoking heroine, that's pretty gnarly. I saw the Grateful Dead I think 28 times before Jerry Garcia died. So, I am also kind of a "Dead-Head" at heart.

I appreciate you even asking me about music, because to much of the public, I am just this one-dimensional figure. For example, recently I shared some photos on social media while I was hiking and someone mentioned, "Don't you mean you are Squatchin not hiking?". Sometimes the public has expectations of me although I am not really interested in those. It turns out that I am just a regular human being with interests.

Where is the favorite place you have visited? Favorites are hard for me because they discount everything else and everything has its own value. I would have to say the Pacific Northwest, that's why I chose to move here. I decided life is too short and we are all doomed to die sooner rather than later, so I wanted to live where I vacationed. The Pacific Northwest is magical. So now it's not visiting, it's just hanging out and being home.

The more exotic locations, Australia was impressive, for Bigfooting stuff in particular.

What is something that most people don't know about you? I don't know what most people do know about me [laughing] so that is a hard one to answer. I am extremely interested in Eastern Philosophy, particularly Taoism, it has influenced my life in several ways.

If you could take one person past or present on an expedition with you, who would it be? Bob Titmus, he passed away too early. I was already Bigfooting when he was alive although we did not correspond. He was the unsung hero of Bigfoot research, even more so than some of the Horsemen in my opinion. I think he deserves more credit, although unfortunately, Bob didn't document much about what he did. Then tragically much of his evidence and records were destroyed when his boat caught on fire and sank in British Columbia. I would have been interested in peppering him with questions and seeing what he knew and what he had learned. Many don't realize the influence that Bob had, although some of the Horsemen did not like him. To my knowledge, Bob is the first researcher to write about tree knocks, he basically discovered it in the late 1950s.

When and why did you get involved in Bigfoot research? I have always been weird and interested in strange stuff. I grew up in the 1970s, I used to watch *In Search Of* and all the monster stuff; crystal skulls, UFOs, ghosts, Bigfoot, and such. My mom was a huge fan of the *Universal* monster movies like the Wolfman, Frankenstein, and Dracula, so I grew up on that type of "black and white" media. I just love monsters. I remember the Wolfman really freaked me out and Bigfoot was kind of "one of those" also and scared the hell out of me. Later, when I was in college, I stumbled across a couple of books and compilations of journal articles written by scientists, including Dr. Krantz's book, *The Scientist Looks at the Sasquatch* and the Halpin and Ames book, *Manlike Monsters on Trial.* I then realized Bigfoot was not only weird and quirky and everything I loved, but it may have also been real!

So, I decided that summer of 1994 to go camping and backpacking in Bluff Creek to see if I could find the PG site. I went and found footprints and I was hooked. That was approximately 28 years ago, one never knows how the wind will push their sails. Back then, I thought I would be teaching elementary school for my entire career and I would have been happy doing so. As I said previously, death is coming for us all so let's make the most of life, and getting into "Bigfoot" ...sure, why not, how fun is that?

What researchers have influenced your work? Grover Krantz and Dr. Meldrum. I never had the opportunity to meet Dr. Krantz, a lifelong regret. I am lucky enough though to call Dr. Meldrum a friend, he stays at the house when he is in town, we go on expeditions, he taught me how to steam eggs [laughing].

What type of eyewitness accounts do you investigate? The things that draw me into witness accounts are the ones where physical evidence has been retrieved (which almost never happens). These types of reports detail a reoccurrence of individuals in the same area. For example, there is an area near the museum that we are working; sightings of a white, black, and red-colored Bigfoot have been documented over the past 20 years. That's an interesting thing because we want to keep track of the individuals. Are they moving together? What are their patterns? Where do they travel at certain times of the year?

If they exist, what do you feel Bigfoot are and why? They are Sasquatches, I feel it is a fault of human thinking that we are so obsessed with this question. Now what are they related to, that is a big question. All of the other labels that we put on them are to make us feel better about categorizing them. At the end of the day, it is a judgment, just like people see me out in the woods and say, "Cliff didn't go hiking he is a Bigfooter", thank you for making me feel superficial [chuckles]. That's what humans do to Sasquatches, it is unfair to them to say they are a human or an ape-like there is a difference. That is like saying, "Is that a dog or a canine?" They are Sasquatches, everything else is our own definition that we are putting on them unfairly. They are certainly apes, but so are you and me. They might be hominins; I am interested in that. Of course, that is going to step on a lot of fundamentalists' and creationists' toes. Some will say that they are Nephilim, Hanuman, or *this* or *that*, or whatever religion you adhere to - there must be an explanation to them because religions are all about explaining things with some sort of belief system. At the end of the day, science is different in that aspect.

What are they related to? I don't think we know enough about Gigantopithecus, to be fair. We know about what they ate, and where they lived when they lived, although we don't know much else. We can only extract limited data from a mandible. I think they are a possible contender, although I am not really in that camp. We know for sure and for a fact that Gigantopithecus was in the line between Sivapithecus and orangutan because of the protein studies they got from the teeth. At the same time, both Sivapithecus and orangutans both lack a brow ridge and Sasquatches are uniformly reported with one. Of course, it could have evolved independently, there is no reason that this couldn't have happened. I am not ruling out Gigantopithecus or Meganthropus, or maybe they are something else that has not yet been discovered in the fossil record. Approximately only 2 or 3% of everything that lived has been discovered.

I am more interested in the idea that Sasquatches might be some type of Australopithecine, specifically the Paranthropus. This theory started to get some real traction in the early 1970s when Gordon Strasenburgh authored a paper hypothesizing that these things may be a Robust Australopithecine which is what a Paranthropus is. I read a fantastic book by Ian Tattersall, who is a paleoanthropologist and an emeritus curator with the American Museum of Natural History in New York City. The book was called *Masters of the Planet*. He wrote about Australopithecines and discussed their probable behavior styles and that sort of thing. Based on his descriptions, to me, they seem indistinguishable from Bigfoot except that they were small. Tattersall also discussed their pelvic structure, which scooped upward to hold their internal organs compared to the human pelvis which faces forward. He also mentioned that this would also give them a somewhat pear shape. At that moment, I said, "holy crap"! I have access to a Sasquatch photo from Oklahoma that sure enough is pear-shaped. I always thought that was peculiar and maybe it was a pelvis attribute. I then went to the *PG Film* and she is pear-shaped, her hips are 2 to 3 inches wider than her chest. I was thinking, "Gosh there is some physical evidence that these things might be Australopithecines or Paranthropus".

Upon more digging, scientists estimate Paranthropus' were about 5 feet tall. We don't have any evidence of Paranthopines outside of Africa. Although if they did radiate out of Africa, they would probably migrate northeast since that is where the land bridge was and how Homo erectus and others got out. If they went in that direction, they would have ended up in the mountains of the Himalayas and further north into Siberia before crossing the land bridge. From what I understand, the most recent Paranthropus fossil is approximately 800 thousand years old. It is also possible that they may have radiated across the land bridge a million years earlier. Potentially as they moved into colder climates, then *Bergman's Rule* would take over. It would not be a huge jump for Paranthropus' size to increase from 5 feet to 8 feet. It would be a small evolutionary change if you have a million or two years to do it. So, I think at this point Paranthropus is a very reasonable guess.

When you look at the Sasquatch behavior, there are a couple of things missing from that behavior that can help us speculate what they may be related to. Firstly, tool use: we have not found evidence of them fashioning tools. They also don't seem to be making and controlling fire. So, when we go back into the Hominin family tree, the first species that we know for sure that did these things was Homo Habilis. So, you would have to think that if Bigfoot don't have those behaviors, we must look before Homo Habilis for the last common ancestor. That gives us the Australopithecine, which is what came just before them.

What excites you about this subject? I love fieldwork, when you are hot on the heels of a Bigfoot and receive a recent report and have a good opportunity to find footprints, trace evidence, or an actual animal. I enjoy being on the scene trying to retrieve something beyond the words of a witness. I collaborate with other researchers throughout the United States and we share data. When I get a new footprint cast from Kentucky for example, and I can compare it to the other footprint casts I received from the same area. Some of those prints from different time periods seem to be the same individual and that's cool.

This is a learning journey for me, I am not trying to prove them, and I am not going to pull the trigger and that is what it is going to take for science to accept them as a species. Maybe along the way, I can help educate the public to soften the blow of discovery when that does happen.

If Bigfoot is a large primate, what environmental conditions would it need to survive? Food, water, cover. They seem to find cover in very strange places; South Dakota for example is mostly plains and the Bigfoot are living in the river bottoms, these deep, nasty, thorny crevices. We don't know much about their metabolism although they must eat a fair amount of food. Most apes have slow metabolisms though, which would enable them to eat less over time. In addition, there is an abundance of meat around; frogs, insects, and carbohydrates; which can be found in lichens and old men's beards, for example. There is a myth that the Pacific Northwest and the North American forests are devoid of food, which simply is not true. In Oregon, there are 30,000 – 35,000 black bears, so there is more than enough food for a handful of Bigfoots.

How would you explain the elusiveness of Bigfoot? It is a combination of different habits than us, we are the ones on the roads and trails, we are active during the day. They live off-road with a *4 x 4* mentality; that mountain, hillside, rhododendron thicket, or any thickly forested whatever is no boundary to them, they just go straight through it. They are so big, so strong, what would be a boundary for us is not one for them. I also feel Bigfoot are just so rare, and they are elusive, their senses are heightened. I think they could be much like mountain gorillas that way. Mountain gorilla troops move through the jungle together so they can listen and keep track of one another, I have evidence suggesting that Bigfoot are like that too. In my opinion, I also think they use river noise, rain, and such as cover. They are very aurally focused, focused on the sounds in an area and use them to their advantage whether it's the quietness or the noise.

They are intelligent as well, although I personally don't think they are as smart as the *Community* seems to elevate them to. They are certainly the smartest thing in the woods, they are not using math or fire of course, Bobo says they are smarter than us because they don't have jobs or pay taxes [laughing]. I feel they are strategic, they work together, there is evidence of that, although they are not infallible, they make mistakes, just like we all do.

I will put it into perspective, people come into the shop every week and say, "they must be just geniuses because they are never seen". I respond, "You grew up in Oregon, how many black bears have you seen here?". They typically say a dozen or so and I explain that there are 30,000 – 35,000 bears that live in Oregon. So, in reality, they have only seen a fraction compared to the population. Furthermore, we are not attributing black bears with having these hyper-intelligent abilities, they are perfectly normal wildlife that are smart.

What type of research are you currently involved with? We have 4 or 5 local areas that we are keeping our eyes on, a few are more productive than others. Currently, we are focused on finding potential footprints to cast. Obviously, we want to encounter a Bigfoot, and maybe try to film it. Apart from possibly getting a Bigfoot on a thermal a few years ago, for the most part, we have failed on filming them. My focus is learning the behaviors of the Bigfoots in the particular areas we are studying. I also want to learn how big those areas are. For example, I am interested if the Bigfoots in Ripple Brook are the same ones over in Fish Creek. That would tell us a little bit about their behaviors. Paul Freeman caught a lot of flak although he was a great researcher. He obtained his footage by keeping track of behaviors like this. For example, he noticed all the activity in Deduct Springs that took place in August and September, so he started driving up there every day. A few weeks later he got footage.

Documenting is critical. If researchers took better notes and recorded better data, we would be so much further along. Also in my opinion, and this may be the old man in me yelling at the kids, "Get off my lawn". It seems many researchers are more interested in having experiences and subjectivity as opposed to gathering actual real data.

What do you feel is your strongest skill/talent that helps you in this field? My strongest skill I think comes from being an elementary school teacher. I can take rather sophisticated concepts and then communicate them effectively to the public who may not be aware of the science or implications of them. My strongest skill for research I suppose is my enthusiasm for getting physical evidence. I have closed the shop early to go get footprints [laughing]. I put "Gone Squatchin" or "Recent Sighting" on the door. It shows our museum visitors who may not feel there is anything to this Bigfoot thing that we are actually doing something.

How did you get involved with Finding Bigfoot? I got a call one day from a guy who was in charge of talent for Discovery Communications which is the parent company of Animal Planet. He said, "We heard you might be good for this. Do you want to go look for Bigfoot on TV?" I said, "Yeah sure, sounds fun, consider me". He asked me to send him footage of myself doing stuff in the woods and they would review it and get back to me. That was November 2009, and I sent him videos I had previously done for my blog and then all went quiet. In May I received a call from Matt Moneymaker telling me that I got the gig. I said, "Great what are you talking about?". He said, "Remember that guy who called you? Well, you got the gig, it's you, me, Bobo, and we are trying to nail down a biologist". I said, "Are you kidding?". I was thrilled and surprised as I had not thought much about it after that initial conversation.

We filmed the pilot after the school year in Alaska. Once that was complete, they sent it to the network and they liked the content and bought 6 episodes which were scheduled for the following winter. That's how it all started for me.

What impact do you feel Finding Bigfoot has had on the Bigfoot Community? Well, I feel it increased the numbers of Bigfoot enthusiasts for sure. My concern was always how are we impacting how the subject was viewed. My goal was, similar to the Hippocratic Oath that doctors take, which is "Do no harm". That has always been my take with the Bigfoot thing. Television is a shallow superficial medium as best, most of it is *BS* and lies, especially when it comes to the more unusual subjects. That was always my concern from the beginning. I felt Sasquatches must be respected through this, we can't produce shlocky *BS* material.

It did get more people interested in the subject and put more people out in the field which was a good thing. Most likely the viewers thought, "All that these idiots (us) are doing is banging on trees and yelling. I can do that".

If I narrow it down to one overarching good thing, it's that the witnesses felt validated. By watching the town halls or our interviews, I feel it helped the thousands of people who have seen Sasquatches and are concerned about sharing it because they don't want to be thought of as crazy or liars, etc... By seeing other people who had similar experiences, that had to help them realize that they were validated in their experience and not crazy or alone.

What is the most compelling thing about foot and handprint evidence? Most wildlife biologists when they are studying mammals, don't rely on observations of the animals, they rely on signs. All mammals are notoriously difficult to photograph, they are nocturnal and most stay as far away from humans as possible. We rely on their spoor to tell us what they are up to; how they forage, their population, their size, their range, etc.... Those are the sort of things we can learn about Sasquatches through looking at their footprints. It has changed my way of thinking about them dramatically. If there is an area which has produced two or three sightings within five to ten miles of each other, they are probably the same animal or at least the same family group. So, the footprints can tell us that. I have a degree in music [laughing] and that is what I know, you talk to a wildlife biologist and they will tell you many more things that the footprints can teach us.

Do you feel since the PG Film that we have gotten closer to proving the existence of Bigfoot? We are just as far from proof now as we have ever been because the proof is a dead one, there is no other way around it. We can speak to the public and in certain ways that will benefit a Sasquatch. The PG film does not do anything to prove it to the public although it helps validate it to the people who are inclined to believe that these things are real. The PG film is solid evidence, it stands up repeatedly, it is absolutely a Sasquatch in the film. The public though is not going to read Dr. Krantz's, Dr. Meldrum's, or Bill Munns's books to find out why this is actually true. Proof is out of my control unless a really small one runs out in front of my car [laughing] and I accidentally hit it, which is unlikely. If it does though, I have a dash cam [laughing]. Actually, almost half of sightings happen on roads while the witnesses are driving.

What are your current projects and how can people follow what you are doing? We have dozens of projects going on at the NABC. We recently obtained a microscope for hair analysis, we have field study areas, and we are going to create mail campaigns targeting areas where there have been sightings. I don't push much out to the public anymore due to the hate I get about what I am doing. We do have a museum membership though and members get access to some of the research we are doing. This way our research is being shared with the people who want to follow it.

I continue to do speaking events, I am trying to be a little less public though. It is good publicity although not good for me personally. It seems that everybody wants to take a piece from me and pretty soon I don't find that there is anything left for Cliff. So, I am trying to take a few steps back, do what I do and share my findings with people who are specifically interested in what I am doing instead of the public. Being a public figure has been exceedingly difficult for me psychologically, [and] my mental state suffers. I am a worrier and I get too stressed out sometimes when I am spread too thin. The public appearances are fun though. They bring some interesting things my way and it is good to see friends from various parts of the country. It also helps to have a little extra income. Since my wife and I are owners of the museum, there are months that we don't take home a paycheck so we can keep the lights on.

What is your elevator pitch for open-minded skeptics? First, I would congratulate them on being skeptical because there is so much *BS* out there. Television is almost all lies when it comes to subjects like these. The internet is terrible and most of YouTube is nonsense. Typically, if you hear about the subject, it is because the individual wants you to hear about *them* and not the animal. I would point out that there is a small amount of compelling evidence and I would encourage them to investigate it. I am not going to do the work for them. I also don't care what other people think, I am not trying to convince anybody. If they do look into it, that shows that they have an interest in the subject.

What suggestions do you have for new researchers? I do feel that any serious student of Sasquatch really needs to delve into paleoanthropology, I think that is a prerequisite. My advice for researchers though is to do it because you enjoy it, don't do it to be successful. Typically, when you go out specifically to look for Bigfoot, you will fail unless you adjust your goals. My advice is to go out to the wilderness specifically where people have seen Bigfoot prior and do something else besides look for them. Go out for another reason; mushroom hunting, fishing, hiking, etc.... If you go out for another reason, you will come back successful most of the time, and maybe Bigfoots show up, maybe they don't, which is most likely the case. This way you will feel more successful and perhaps more fulfilled. Be prepared for the Bigfoot thing because, at the end of the day, Bigfooting is essentially camping with a purpose. Most people who see them are campers, so go camping. Being outdoors is good for the soul, body, and mind. Look for them, and at the same time, you are going to discover a lot of cool things about other animals, plants, and insects, etc... The more you learn about the environment, the better Bigfooter you are going to be. Enjoy the environment, it is impressive and Bigfoot are one of those amazing things that are out there in it.

Summary: My concern is them, the Sasquatches, I just don't care about the people. There are many individuals I like, although people as a whole are not my concern. I think far too many Bigfooters make it about themselves instead of the animals. They may say, "Look at me, look at what I am doing". People lose sight of the animals which seems to be the point of us doing this. I don't lean toward the narcissistic side that I feel many others do in this field; they want to be "somebody". Maybe that is a thing of the youth, for me although, it is an easily discarded mental quirk as I grow older.

How to follow Cliff Barackman:
cliffbarackman.comnorthamericanbigfootcenter.com

LYLE BLACKBURN

Lyle Blackburn is a native Texan known for his work in writing, music, and film. He is the author of several acclaimed books, including *The Beast of Boggy Creek* and *Sinister Swamps*, whose subject matter reflects his life-long fascination with legendary creatures and strange phenomena. Lyle is also founder of the rock band, *Ghoultown*, and narrator/producer of documentary films such as *The Mothman of Point Pleasant* and *Boggy Creek Monster.*

Lyle is a frequent guest on radio programs such as *Coast To Coast AM* and has been featured on various television shows airing on *Animal Planet, Destination America, Travel, Discovery Science*, and *Shudder*. In his work with Monsters and Mysteries in America, he served as both consulting producer and special episode host.

As a musician, Lyle has achieved similar success. His band *Ghoultown* has released nine albums, which have not only earned a loyal worldwide following but found their way into movies, video games, and numerous live venues across the United States, Canada, and Europe. Highlights include an invitation to write a song for iconic horror maven, Elvira Mistress of the Dark, which aired on her nationally syndicated show, *Movie Macabre*.

When Lyle isn't writing books, hunting monsters, or performing with his band, he can be found speaking at various cryptozoology conferences and horror conventions around the United States. Just look for the trademark black cowboy hat.

Knowing his books, podcasts, and documentary work, I met Lyle in person for the first time at the Ohio Bigfoot Conference in 2021. He was very well-spoken, informative and captured the audience when presenting. This is also evident in his books, where he shares extensive historical documentation and research while providing in-depth accounts of the Bigfoot phenomenon in the United States. Working with Lyle for this book, I also found him to be thorough, prompt, eager to help and easy to talk to. He also shared a few hilarious stories! Lyle has a wide professional and personal background which certainly makes a positive impact on the projects he is involved in. His work on the subject has been influential in the Bigfoot community and I look forward to future projects to come.

Where did you grow up? I was born in Fort Worth, Texas and have lived in the area my entire life.

Hobbies outside of BF? I collect monster movie memorabilia and I build monster models. Work to me doesn't feel like work, my hobbies bleed into almost everything I do, it's like being a kid, I enjoy it all, it's fun!

Favorite musicians or bands? My favorite bands are *The Misfits* and *Mercyful Fate*, I have been a lifelong fan of *Kiss*. I enjoy everything from Metal, Punk, Old Country, a little bit of everything.

Most memorable event and where was it? Oh boy, there are many great concerts I went to. I saw *Metallica* on their *Ride to Lightning Tour* which was their 2nd album at a small-sized venue which was really cool. I have seen *Danzig* quite a few times, I wasn't old enough to see him when he played with the *Misfits* prior to them breaking up in 1983. I did see him with his band, *Samhain* and one of those cases in a ridiculously small club the police arrived and shut the show down. That stands out as a memory although they didn't finish the show.

Favorite Place you have visited? My favorite place is Fouke, Arkansas, the home of the *Legend of Boggy Creek*. It has played such a huge role in my life, especially after having written the books and it has been a focal point of my life for the past decade. Going to that area, spending time in the bottom lands, the woods, and getting to know people that were associated with that famous Bigfoot case, it is just my favorite place in the world.

Favorite childhood memory? I would say spending time outdoors, riding bikes with my friends. From early morning until sundown, we would just leave and we had that freedom back then of being able to go and your parents would just say be home before dark. We would play in the woods and explore. Those memories of the freedom, the summertime and the countless hours with my friends are something that I remember to this day. One of those friends is in my band *Ghoultown,* we have spent 21 years together touring around the world and still having fun.

What is one thing about your childhood that you look back and say, "what was I thinking?"? I think I was in 7th grade, my school allowed us to dress up for Halloween. I was a huge *Kiss* fan and decided to deck out as Gene Simmons. I had a wig from my mom and a previous costume, I performed in elementary school in talent shows lip-syncing *KISS.* I also borrowed my mom's leather platform boots, which I really didn't think that through because walking around all day was brutal, I remember thinking; "dude this is ridiculous and stupid". I actually took them off at one point so I was a Gene Simmons in socks (laughing) and just looked pretty dumb.

What is your favorite seasoning for food? Well, it would have to be my *Monster Sauce* hot sauce, I am obligated to say that (laughing). It truly is good, I literally love that stuff, it is always here in the kitchen and we put it on everything!

What is something that most people do not know about you? I'm a self-trained chef.

If you could take one person past or present on an expedition with you, who would it be and why? Adam Davies because he's knowledgeable, skilled, and funny!

When and why did you get involved in Bigfoot research? As a child, the one thing that had the most profound impact on me related to the research and work that I do now was seeing the movie *The Legend of Boggy Creek*, at a drive-in theatre. That stuck with me, and it catapulted me into doing research and writing my first book.

As far back as I can remember, I loved monster movies and related things. In elementary school, I received a *Scholastic Reader* that had stories of Bigfoot, the Yeti, and the Loch Ness Monster. When I read those, I remember saying "wow there are actually real-life monsters out there". Then after seeing the movie, that did it in, since Boggy Creek was only 3 hours from where I lived, I knew there were monsters geographically close to me.

If you have had an experience(s), what is/are the most compelling one(s)? I have not had a personal sighting that I definitively say was Bigfoot or another cryptid, however, I have had experiences where I think I was close. I was in Boggy Creek and something howled several times and my research partner and I. We felt that the sounds were something that we couldn't identify and possibly what people would say is the creature. We were canoeing along a bayou late that night and then traveled back down the bayou channel to our camp. Whatever this creature was that we heard previously, must have followed us and howled again from across the channel. I grabbed my flashlight and plunged down the hill and the creature took off and howled again from approximately 50 yards away.

What was the most compelling witness account that you heard first-hand? There are quite a few, there are so many reports and many people that seem to be credible. Then there are a few that you just can't explain. There was one report where 2 female witnesses in Florida had seen something in broad daylight. They were driving in an area where there have been Skunk Ape reports. The road was lined with trees on either side and as they came to a bend in the road, they saw an upright, hairy, bipedal creature walk across the road in front of their car. They felt it was an animal and not a potential hoaxer since it was a dangerous area to cross the road with the curve and high-speed traffic.

I interviewed both witnesses in person, they also had a video they recorded immediately after the incident. They gave me the video, I heard their real reactions, this is something that they absolutely experienced and saw. It is rare to get multiple witnesses at the same sighting and with noticeably clear daytime circumstances. I can't say what it is they saw; I can guarantee though that they saw something unexplainable.

What do you feel Bigfoot/Sasquatch are and why? An undiscovered species of ape.

What excites you about this subject? To me, the exciting part is uncovering cases where people have seen unexplainable creatures and where I can also find historical accounts and modern-day witnesses. Putting these two things together for me has substance. Those types of occurrences have built the foundation for several of my books; where there are significant cases and have even transformed small towns. I love those the best.

If Bigfoot is a large primate, what environmental conditions would it need to survive? Forestry, water, and seclusion.

How would you explain the elusiveness of Bigfoot? There are very few of them. (Less than people seem to think.)

What are your current research goals and how do you go about them? To continue to document interesting cases and offer engaging books. I have been researching and collecting encounters with various entities and cryptids that don't necessarily fit into an atypical single book format. Most of my books cover one specific case or creature. I want to develop a book based on all these random encounters. I have gotten many cool reports over the years and would like to share them.

What do you feel is your strongest skill/talent that helps you in this field? I feel the fact that I am a professional writer, I have a degree in English and was always good at it. I feel that I address the subject from more of a journalistic standpoint who also loves this subject. According to my readers, the books are professionally written, exciting, and hard to put down. My approach may differ from someone who has an experience and then authors a book, which many are well done. Once I research the subjects, I feel I can tell the story in an engaging way.

If you could get better at one research-related skill, what would it be and why? I feel that I have a good process in place now, I suppose I could get better at seeing the creature when I am following up on a report (laughing). That would be an improvement, I could investigate a case and actually see what the witnesses saw. That is the key to the elusive part of authoring these books, much of the accounts are hearsay, anecdotal reports, and conjecture rather than a true biological field study of an animal for example. My own sighting would be icing on the cake for me so I can give a personal description and cap off the book related to the occurrences.

What is your biggest accomplishment in this field? Having thoroughly documented the history of the Fouke Monster (of The Legend of Boggy Creek fame).

What is your elevator pitch for open-minded skeptics? My usual approach is to try to identify with them first and let them know that I understand that this may sound fantastic and impossible and agree that we don't have a body to examine. So, it's understandable to be skeptical. I then would share a few cases of compelling reports. Also share that there is possible evidence, footprints, and unidentified hair samples. The average person does not realize the depth of possible evidence and eyewitness accounts.

What are your must-read books for skeptics and new researchers? One of the classic and best books was written by John Green, *Sasquatch The Apes Among Us*, I modeled my book writing after this. Also, in *The Essential Guide to Bigfoot* by Ken Gerhard, he explains the aspects of Bigfoot very well in modern terms.

What suggestions do you have for researchers who want to get involved in this field or further educate themselves? The best way to get up to speed on Bigfoot is to read books. I constantly see people debating about things they see on television or social media. Sometimes those are not reliable sources, none of it is as well researched and as well explained as you can do in a book. Sometimes I see debates on social media over a photo where the individuals do not realize that the photo was de-bunked years prior. First, go to the source of books and understand the background and some of the science around Bigfoot. If you decide to join social media groups, investigate them thoroughly and find the best ones. Furthermore, physically go to the areas that you are interested in researching, interview people, do some fieldwork so you can get your own perspective instead of just relying on television and social media.

Why do you think there is not more evidence or proof? I think if the creatures exist, they are far rarer than they appear to be, because of the plethora of television shows and daily social media posts, it gives the appearance that these creatures are running wild all over the place. It also gives the appearance that researchers will go out one weekend and have an experience. Due to their rarity, it makes it hard to find evidence. When we are in the woods, how many times do we come across a dead bear or bones of one? We don't, and bears have a much higher population than a Bigfoot. Possibly for every 25,000 bears in a specific area, there may only be a population of approximately 300 Bigfoot. This is an example of why it is extremely hard to find evidence.

How do you feel mainstream scientists will publicly accept the existence of Bigfoot? The only way mainstream science is going to accept the existence of Bigfoot is if there is a type specimen, a body. At this point, Bigfoot has become so much a part of pop culture and commercialized that it is understandable why anthropologists and mainstream scientists are not going to put much time into researching it, partly due to possible ridicule from colleagues. I am not advocating anyone to go kill one just to prove it though, it's not my agenda. No matter how good a photo or video is, without a body it won't be accepted.

How to follow Lyle Blackburn:
lyleblackburn.com

STACY BROWN JR.

Stacy Brown Jr. has been fascinated with the Bigfoot Phenomena since he was a young child and has been actively investigating for over a decade. In 2012, while camping in Florida, he and his father captured what many researchers and enthusiasts feel is one of the most compelling thermal videos in the existence of a potential Bigfoot.

Stacy is currently a film director and editor, and he was one of the winners of Spike TV's *10 Million Dollar Bigfoot Bounty* in 2014. In addition, he has appeared on *Finding Bigfoot* and has been featured on multiple media outlets and podcasts. He is also a member of *Outkast Paranormal*, where the group experiments and researches all things unexplained.

Brown Footage

Stacy was extremely candid, and open during our conversations and interview. I appreciated that he did not hold back, and his passion for the subject was clear as he shared his thoughts and theories. I enjoyed going back and listening to our interview while writing this, it was certainly entertaining and informative. I respect Stacy's perspective on the subject of Bigfoot. The video he and his father captured continues to intrigue researchers and offers more potential evidence that the elusive creature exists.

Where did you grow up? Crawfordville, Florida. It's in Wakulla County. It's a super small town, growing up, most of the roads were dirt and there's literally woods everywhere.

If you could take one person past or present on an expedition with you, who would it be? I would take my dad just so I can see him again. That's for different reasons other than Bigfoot, I could just hang out with him again.

Was it your dad who got you into this? He was a fan *of Unsolved Mysteries, In Search Of,* and things like that. So anytime he was watching TV it was either football or shows about mysteries. That's what my dad loved and I watched with him and ended up loving both of those things, paranormal and football. As a kid, because we lived in just such an isolated place, I thought Bigfoot was literally in the neighborhood. Turns out he might have been, I ended up years later getting reports from that neighborhood. I thought, "What the hell?".

Watching those shows with him kind of got it started and moved the ball forward. As we got older, we were doing it as a way just to hang out and spend time together. We both loved being in the woods anyway, so, we would look for Bigfoot. Then we started finding things that made us question a possible existence of the unknown creature.

What excites you about the subject? The thing that excites me is I know it's real from my own experiences. I only focus on my research; I've successfully been able to shut down whatever everybody else has got going on in the world of Bigfoot. I don't hear that noise anymore. Some of the theories don't make sense to me and it's gotten politicized for the best part of it.

You have people who are in this ape camp, good friends of mine, like Cliff Barackman. Then you get other people who are in the "Woo" camp, and hey, I got friends that are in that camp also. Hell, I'm in the camp of, what if it's all of it? I don't know what the hell it is. Early on I just wanted to prove it and to leave my legacy behind. I thought maybe I would get a sign or a statue of myself somewhere [laughing]. That's what I wanted, I know it may sound f**ked up, but that's how I was. Now, I don't think I'll ever know, I don't think we will understand until we grasp physics better. It may take another 100 years for us to figure out exactly what the Bigfoots are doing.

Do you have any theories of what Bigfoot are? I absolutely have no idea what they are. Could they be just a normal primate like half the Bigfoot community thinks? Sure, could be that. What I saw the two times I had sightings, that's what it looked like, so it's very plausible. However, why are they not in a zoo yet? There are people out there who are excellent hunters, they do it for a living, you know what I'm saying?

We guess that they must be super intelligent since they are so elusive. I think there's something else to them, I think there's something with the lights, there's something that we're not grasping. On the other hand, there are only a limited amount of people that are looking for Bigfoot. So, they may end up being just a regular primate and not that elusive. I do know that when I look for them there's a reason my equipment goes dead, the lights show up, the tracks disappear and my cameras turn off, it is weird.

Do you feel academia should take this more seriously? We are victims of the scientific community. I have a major issue with it. If you say scientists in my house, that is a derogatory term and you may get slapped [laughing]. Hell, they studied sex drives in pigeons who were on cocaine. That's a f**king real story. The government spent money on that.

Pigeon Study

How many witnesses have you interviewed? 400 – 500, I've talked to NASA employees, sheriff's deputies, park rangers, all types of people, even a homeless guy who said he's seen one.

Are there any accounts that stick out to you? I had one where a witness from Ocala saw a Bigfoot that had one arm in the sleeve of a flannel shirt. That sticks out because it was just odd. If that was the case, then that Bigfoot was watching people and mimicking, which is very cool to me.

Another was a NASA scientist who saw a Bigfoot in the daytime sighting while fishing. As he rounded a corner in the creek this thing started throwing stuff at him from the shoreline.

When it comes to witnesses, I land somewhere with around nine out of every ten people, for whatever reason, are bullsh*tting, lying, or misidentifying what they experienced. There are many people who are genuine and then there are several who just want to have an encounter with something, I guess, to make their life different, I don't understand it.

Can you share what happened when you had your 2012 sighting? My dad had just beat cancer and wanted to get back out in the woods again. We decided to go to Torreya State Park and set up at Rock Creek Campground. We had light rain and wind throughout the day. Once we got camp set up, we started cooking ribs and listening to music. It was already dark and at one point, we heard what sounded like a "tree knock", although Dad figured it was just a couple trees bumping together due to the wind. Then suddenly, we hear a knock followed by a grunt!

We jump up, Dad grabs his *Flir* which he sold his boat to buy it earlier that year. Surrounding the campground were palmettos and they were tall and provided plenty of cover for something to hide. Additionally, the cicadas were extremely loud, and deafening, and beyond the light of the campfire, it was pitch black. We head towards where we thought the grunt came from, the wet ground helped our stealth since the leaves were not crunching as we walked. With the distraction of the cicadas, campfire and music, the whole thing was like a perfect storm for if you're trying to sneak up on a Sasquatch.

As we approached, it sounded like there were 3 or 4 of them, they were just running and f**king around, I don't know what they're doing, maybe hunting. We get to a game trail and we hear something coming down the ridge. It suddenly stopped, we assumed it detected us. Dad throws up the *Flir* where we think it is. Initially, he sees heat signatures and movement and thinks that it is two raccoons fighting because he sees images on both sides of the tree. Next, it darts out from behind the tree and runs deeper into the woods.

Dad could have had a fit, bro, he flipped out. He turned around and said, "Get your gun, get your gun". I heard it run, I just didn't know what the hell it was. We got back to camp and could still hear them making noise. I wish we could have done some things differently, like stayed, but my dad wasn't having it. He became a believer that night, it was cool, man.

What characteristics of Bigfoot are different between the reports in Florida versus the Pacific Northwest? I think that it is a cultural difference between the people. In the South, people carry guns and feel threatened by wildlife and the potential of Bigfoot. In Washington State you have more hippies, so you have a nicer Bigfoot. I think we're dealing with the same thing; I don't think these things are safe to be around at all. I think the government is aware of where most of them are at and they try to avoid letting people in those areas for their safety. Let's say, it's just an ape, you know how aggressive chimpanzees can be. Are you going to go into a chimpanzee enclosure? Hell, no, you ain't.

Take a black bear, for the most part, is going to run away because they don't want to have anything to do with you. Although, it would still tear your head off if it had to. You're not going to go to sleep with food in your tent, right?

Do you think the physical traits are a result of the environment? Yes, it makes more sense. The one I saw here in Florida looked like it lived in the heat. The one I saw in California looked like it would have died if it had lived here, too big, too furry. Nothing like that you can survive 100-degree, 95% humidity days, not with that much hair on your body. I think it's like the white-tailed deer here versus the white-tailed deer in Idaho. You also have just the cougar or puma, one is smaller and closer to the equator, some refer to this as *Bergmann's Rule.*

We may also have a different creature in southern Florida according to the reports. I live in the part of the state where we have Bigfoot reports. Once you get to Ocala and south you have skunk ape reports. There's a difference in the creature, witnesses report that they look like that orangutan-type photo the Sarasota Police Department put out.

What advice do you have for a new researcher? First, I'd tell them not to read any books. I'd tell them to go camping and write their own book. Reading some of that bulls*it from the groups out there, like the little pamphlet they give when you join the group, you know what I'm talking about? That stunted my growth as a Bigfooter for the longest time. I'm going out there hollering at the top of my lungs, trying to find something to call back to me and I'm beating the trees. The best thing to do is to just shut your mouth and go to the woods and just take it in. If you're just out there, they'll communicate with each other. Do you know anyone who hunts by going out and hollering in the woods? We were walking around in large groups of 20 people at times. Everybody had lamps on and we were doing the most peculiar sh*t in the woods, in the middle of the night. I wonder what Bigfoot thinks when he sees these large ass groups of people just strolling down the road in the middle of the night. Just go out there and figure it out for yourself.

What is your elevator pitch for open-minded skeptics? I know it sounds crazy, the thought of looking for Bigfoot. I know what I saw though and I'm not lying to you. Why can't they exist? How many people are in the woods looking for Bigfoot anyway? Today, people are going outdoors less and less. I talked to a family member who was adamant about not believing in Bigfoot and yet turns around and puts faith in religion. I mean Jesus didn't leave tracks outside my tent, bro.

How to follow Stacy Brown Jr.:
outkastparanormal.com
Facebook: BigfootStacy
Instagram: bigfootstacy

AMY BUE

Where did you grow up? I grew up in Poland, Ohio. It's a suburb of Youngstown: a really small town.

What are your hobbies? When I'm not teaching or writing grants or doing Bigfoot stuff? I am a writer, so I like to write about things other than Bigfoot, although I'm writing a book that does have to do with it. I'll try just about anything once. I took square dancing lessons and I was really bad at it, but I did it. I like to try new things. I like to hike, and I'm an Ohio Certified Volunteer Naturalist. I volunteer at local parks.

What are your favorite musicians or bands? I like anything. I really do. My favorite is bluegrass and I grew up playing some different instruments, but poorly [laughing]. I have a banjo that I'm trying to teach myself. I also love classical music. My parents took me to every different type of music show there was. Polka, Big Band, you name it.

I grew up loving U2 when I was in college and 80s music. My favorite singer ever was Johnny Cash: love him. I like all kinds of music, really. I love the Ramones.

What was an embarrassing or funny memory while researching or any moment in your past? Oh, my goodness. I have a lot of embarrassing stories. A funny one is when I was up in Stevens County, Washington filming a documentary with Extreme Expeditions Northwest. I was the only woman, and so I was aware of all of the cameras set up on a perimeter around where we were camping. I would normally walk down the road to an old, abandoned outhouse rather than be caught on camera with my pants down, so to speak. I had to carry my own toilet paper, but on one trip, I forgot it. Let's just say that one of my socks was sacrificed that day.

If you could take one person past or present on an expedition with you, who would it be and why? Roland Welker from the History Channel's "Alone: Season 7". He is a beast in the woods, so I would feel safe with him. I could concentrate on Bigfoot, rather than on just staying alive.

When and why did you get involved in Bigfoot research? In 2012, I had a possible sighting and I say possible because it was too far away for me to be sure. I was a passenger in a car going over Meander Reservoir on a bridge in Mahoning County, Ohio. I was looking out the window and there were some birds flying around on a cement block that's out there in the water. That caught my attention, and I saw something or someone standing along the shoreline holding on to a branch of a tree. It was uniform in color and looked to be exceptionally large and wide. I yelled out, "I think I just saw Bigfoot". I immediately thought that was stupid. I'm not sure how fast we were going as that depends on the traffic over that bridge. You can't stop to take a better look. The driver didn't see it, and I was just kind of stunned looking at it.

If they exist, what do you feel Bigfoot/Sasquatch are and why? I don't claim to know. I'm not even 100% convinced that they exist. I have friends with diverse types of theories. My own interest lies in the direction of them being a type of primate. Whether more chimp or more human-like? Not sure.

How would you explain the elusiveness of Bigfoot? My friend pointed me to the Cross River gorillas and how they managed to stay so elusive to Western science for so long. The indigenous people in the Cross River Region of Africa knew that these animals existed. Most Westerners didn't believe them until we finally saw them for ourselves. I think the same thing could be happening in North America. Our Native Americans and First Nations people claim that Sasquatch exists. Maybe we should listen to them. If they exist, they could stay elusive because they are intelligent. I think that if they exist, their numbers are probably small and they stick to areas not frequented by humans. It isn't really surprising to me that they are hard to find.

Is there one of those witness accounts that really stand out to you of all the folks you've interviewed? I was taking reports at a hunting and fishing show, and an older gentleman approached me and my friend Tina Sams. He was hesitant to talk at first. When I first started doing those shows, I wondered if anybody was going to talk to me or if everybody would just laugh at me. I didn't really care if people would laugh at me, but I just didn't want to waste my time. That has not been the case. There were people every day that would tell a story. Anyway, this gentleman finally said, "Listen, I need to tell you something because I've never told anybody this, with the exception of one person". He was at his family farm and he was looking out the window and said that without any shadow of a doubt, he saw a Sasquatch standing in the field. It didn't do anything to him; it was outside and he was inside, and it didn't make any threatening moves or anything like that.

He said he got a long look at it and that it was not a bear, and it was not a person. It was 100 percent a Sasquatch, which he never believed in before. The thing that really stood out to me was this guy was getting very emotional as he was telling the story. Then he said, "You know, ladies, the opening day of deer season used to be my Christmas. My whole life since I was a little boy, I would go out hunting and it was my favorite thing to do. I would come to these shows and buy new things. I still come to the shows, and I still buy new equipment, but since that day I've never been out in the woods hunting again". He said he just will not go hunting anymore, and he will not go out in those woods anymore because he said there was something that shouldn't have been there. But it was there, and it wasn't supposed to be real. He saw it and was shaken up. Those are the ones that really get to me.

What is Project Zoobook? Project Zoobook is a group I co-founded and coordinate. It is formed of primate zookeepers, primatologists, wildlife biologists, marine biologists, forestry workers, archaeologists, anthropologists, university professors, law enforcement officials, taxonomists, and other scientists working alongside Bigfoot researchers from across the country. Project Zoobook meets virtually bi-weekly as a think tank where these individuals discuss research, ideas, and new findings that are pertinent to both the Bigfoot topic and primate behavior. They are collaborating on micro studies of areas currently being researched across North America and on scientific endeavors that could help that research.

Project Zoobook's wish is to carry on researcher Dr. John Bindernagel's work of making the subject of Bigfoot less taboo. Amy was given the International Bigfoot Conference's *2018 Dedicated Researcher Award* for her work with this group. If you are a scientist and are interested in learning more about Project Zoobook, contact Amy.

What would be your elevator pitch for open-minded skeptics? First, I would say that there's more scientific research going on than they probably know about. I will always bring up the Olympic Project's research because I'm just such a fan of how they go about things. I'm in the Olympic Project now, which I still can't believe. The team members who are out in the field all of the time are doing amazing things. I never knew that anything like that was going on before I looked more closely into Sasquatch. I never knew that there was such a vast collection of footprints out there that really don't seem like they are hoaxed. If this person caught in the elevator would talk to someone like Dr. Meldrum, that might convince them that it's possibly a living species that is making the footprints. Then there are the Native American stories: I have been told by ladies from the Blackfeet Nation in Montana that they aren't just moral tales to scare the kids. What many really seem to be saying is that these are real creatures and they're still out there. I would tell them about how different primate subspecies have only been documented by Westerners relatively recently, even though the native people of Africa and Asia have always known they were there.

There are varied reasons why I think that they could be real. Dr. Jane Goodall doesn't discount the possibility. If she can be open-minded about it, then I certainly can, and so could the person in the elevator.

How to follow Amy Bue:
Bigfootamy@gmail.com
Facebook: Amy Bue
Instagram: mrsbuedizzle

PETER BYRNE

Peter Byrne is one of the *Four Horsemen* of Bigfoot and a pioneer on researching the subject. Born in Dublin, Ireland in 1925, his interest in the subject began when he was a child, hearing stories from his father about the Abominable Snowman, otherwise known as the Yeti. At the age of 18, he enlisted in the British Royal Air Force during World War II. He was stationed in India when the war ended. While awaiting his departure back home, Peter had the opportunity to take his first expedition in search of the Yeti. He was hooked, and soon connected with Tom Slick, led expeditions for him in the Himalayas and eventually headed his first project in North America funded by Tom in search of Bigfoot.

Peter is one of the few individuals in history to obtain funding to lead long-term research projects related to Bigfoot. It is no different today, the biggest challenges researchers encounter are time and funding, as academia and science for the most part will not support projects related to Bigfoot. It also was no secret that the *Four Horsemen* and other past researchers had their differences, although if you dig into the history, you will soon understand the impact Peter has had.

More than 60 years later, Peter continues to research the subject and has a burning desire to solve the mystery. Talking to Peter, he continues to be charming and eloquent with a witty sense of humor. My impression from our conversations is that he takes a focused and professional approach to his research. This book would not be complete without Peter Byrne and it has been a special honor to have the opportunity to collaborate with him and have him included.

If they do exist, what do you feel Bigfoot are and why? First of all, I do think they do exist, but I don't think there are as many as there used to be based upon the reduction of credible sightings and footprints that have been reported in the recent years.

I feel we are looking at a hominid or some type of large, hair-covered, upright walking bipedal animal. It is possibly an ancestor of Gigantopithecus, first discovered in 1935 by a Dutch-German paleoanthropologist Ralph von Koenigswald, who at the time was living in Hong Kong.

When and why did you get involved in Bigfoot research? January 1960, I came to the United States. I was previously in the Himalayas for 3 years conducting research sponsored by Tom Slick. At the end of 1959, myself and my associates arrived in Katmandu. We typically hiked back down to Katmandu every Christmas for a week and then hiked back to our research area. There was a cable from Tom Slick which said essentially, "You have been up there long enough, I would like to offer you something else. Would you be interested in coming to the United States and helping me search for a creature there known as the Bigfoot?". I have to say my brother and I laughed, because to us the United States was all freeways and skyscrapers, we had never been there. We found out later that the Pacific Northwest is actually larger than the Himalayas.

We worked on the project for 2 and a half years until Tom tragically died in an airplane crash. The project came to an end and I went back to Nepal.

What is the most compelling evidence that you have come across? Well first, it was the history of the subject, I was able to trace it back to 1785 with the Native North American accounts. I found 5 sets of footprints over the past 56 years that were compelling. Many were large and in remote areas which in my opinion, were made by a large bi-pedal creature.

Then the 1967 footage came along which was remarkable. We actually paid a company in Peoria $85,000 to review the footage, they spent 6 months examining it. In summary, they stated that they did not know what it was although it was a real living creature.

What were the most compelling witness accounts that you heard first-hand? The first one that comes to mind - I had 2 reports from the same location, just a day apart. The first report, a gentleman was working on a tractor. As he was navigating it, he looked up and saw what he thought was a man at the edge of the forest. The figure's arm was resting against a tree and was watching the witness. He stopped and turned off the engine. He then noticed this was no man and this creature was covered in hair and enormous, then the thing walked away and back into the forest.

The next day, 2 men were working in the same area. They had taken a break around 11am and walked into the shade to rest and take a smoke. As they walked towards the shade, what they describe as a creature, stands up and walks away from them. One of the men chased after it and threw a rock at it. The creature did not stop and continued to walk away. All these eyewitnesses were experienced outdoorsmen, familiar with the forest and were credible observers.

In June of 2019, we had another report that we called the *7-man Sighting*. There were 7 loggers working with heavy machinery. These men were professional, each with many years of logging experience.

The area that they were working in was gated like many others in that region. As they are working, one of the loggers says, "who is that walking up the road?". All 7 of the men working that day experienced this sighting. The first visual of the creature was from approximately 300 yards away, and it continued to walk towards them until it was approximately 100 yards away. They stood there with their mouths open and in disbelief, startled, they did not capture any photos with their phones, and then the creature turned and walked away. They confirmed it appeared to be what they referred to as a Sasquatch.

Why don't more witnesses get photos or video? Witnesses are in shock; they cannot believe what they are seeing. By the time they pull their cameras out, the creature is gone. I talked to a state trooper with over 20 years of experience who had a sighting. He was working mainly in the forest regions, helping to reduce timber thefts. He was driving one night and he sees what he believed to be a Bigfoot standing on the side of the road. As his headlights shined upon it, the state trooper hit his brakes and stopped. He initially thought it to be a person and figured if someone is standing on the side of the road at 1 o'clock in the morning they either needed help or they were up to mischief. The creature then walked off, the trooper was in shock, he said it was "like a creature from another world". Shaken by what he just saw, he did not investigate and put his foot on the accelerator and took off. He did go back the next morning with another trooper and found footprints in a bank where he witnessed the creature the morning prior.

Seeing one of these things, especially at a close distance, can be very shocking, if it is a mature creature, you are looking at something that could be 7 feet tall, more than 400 lbs., and covered in hair. It is like seeing a gorilla for the first time when you are not expecting it.

How would you explain the elusiveness of Bigfoot? The size of their potential habitat, they have enormous cover. Many small plane pilots have unsuccessfully attempted to spot animals by air in the Pacific Northwest forests. They are too dense, it is impossible. For example,

Northern California, Oregon, to the Alaskan border have millions of acres of habitat which can support Bigfoot. I am quite certain that there is enough space for a small group of them to survive and hide.

What has kept you going throughout the years? All the fascination of the find. First, with the Yeti, we thought there might be a few of them left at that time I was researching in Nepal. We felt we could find one so we hiked great distances to search for them. Nepal at that time had no roads, no air service, no helicopters, just trails, nothing else. We would travel approximately 20 miles a day. If we received a report of something that happened 100 miles away, for example, it would take us 5 days to get there. The same thing with Bigfoot, when we followed up on reports, we would spend days and nights watching areas with the hopes of seeing one. In the beginning, we carried rifles with us because we did not know what we might encounter, after that we only carried our cameras.

Over the years, I interviewed hundreds of people who had sightings. I created a questionnaire and would use that to help assess the credibility of the witnesses. Many of the credible witnesses were state police, US Foresters, and researchers; these were people who can accurately give descriptions of what they encountered. They were trained to know the environment and report on an incident based on the facts that were present. An interesting recurrence was many witnesses lived thousands of miles apart although shared similar descriptions and attributes of what they experienced, which was compelling.

Have you changed your research techniques as you became more experienced? Yes, later in my research, I have spent less time in the forest. I have learned to rely on technology and eyewitness reports. Originally, we would camp for multiple weeks and months in the Pacific Northwest and British Columbia after we received a report. We realized that this was not productive and learned to become more efficient.

What is your elevator pitch for open-minded skeptics on the possible existence of Bigfoot? I meet skeptics and they typically have not researched the subject at all and have no interest in doing so. They dismiss it with a wave of the hand so to speak. "How could something like that exist?" If they would do a little reading, they would find out that the Pacific Northwest for instance is bigger than the Himalayas and also discover that the basic essential things that animals need to survive are present. I would share with them that a Bigfoot, like a bear may be an omnivore, eating anything that was available. There is plenty of food available. In the Pacific Northwest, we have approx. 100 edible plants, berries, and animals, which would support a large omnivore.

I would then suggest if they were interested, investigate the historical documentation on the subject.

What advice do you have for new researchers that want to get involved in this subject? My advice is to get into the woods, investigate, and use technology, including motion sensor cameras. It is much easier now to place a camera in the woods and leave it versus in the past when we would have to sit in the tree for hours and days and observe with the camera.

Check your cameras on a regular basis, and collect and document all your findings. When researching witness sightings, create a questionnaire to assess the credibility of the witnesses as well as document what they experienced as best you can.

How do you feel mainstream scientists will publicly accept the existence of Bigfoot/Sasquatch? A physical specimen, hopefully, we find one that has died naturally in its habitat.

How to follow Peter Byrne:
www.Petercbyrne.com

LOREN COLEMAN

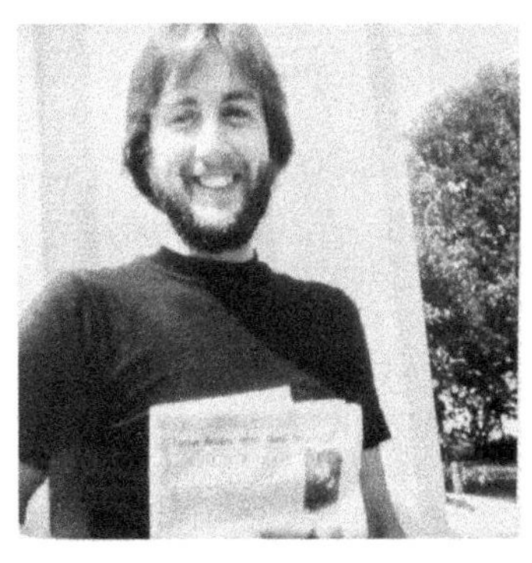

Loren Coleman has been a historical and impactful Bigfoot and cryptozoological researcher since the year of 1960. He may have the most extensive history and experience in the field of cryptozoology. He has authored many articles, published multiple books, and has appeared and been involved with radio and television programs beginning in 1969, with articles about his work published before that. You can keep up to date on anything cryptozoological-related on Loren's blog site, the *CryptoZooNews* which he started in the 1990s. Studied as a polymath and often writing on a variety of topics, in 1987, he wrote *Suicide Clusters*, a nationally recognized book looking at the growing problem with teenage suicides, which was followed by *The Copycat Effect* in 2004.

In 2003, Loren founded and is the current director of the *International Cryptozoology Museum* which is in Portland and Bangor, Maine. The museum includes countless historical exhibits and plays an essential role in the preservation of artifacts, specimens, documents, and one-of-a-kind items. The main museum is remarkable, a must-visit for anyone interested in Bigfoot, cryptozoology, or the general history of living creatures. Visiting Loren at the museum was certainly one of the highlights of this book project. I found him to be gracious with his time and we were given the coolest personal tour. Loren was welcoming, kind, knowledgeable and funny. My wife, Dana, and I genuinely enjoyed our time with him.

Loren earned his undergraduate degree from Southern Illinois University-Carbondale where he studied anthropology and zoology. He then went on to receive a graduate degree from Simmons College in Boston with the primary focus of psychiatric social work. Later Loren participated in Ph.D. coursework at Brandeis University, studying social anthropology, and studied in the doctoral program at the University of New Hampshire's Family Research Laboratory.

Loren's thoroughness and diligence are critical and evident in the success he has had in his research. This certainly has kept me on my toes while authoring this book. Most importantly though, Loren has supported and provided positive constructive guidance on this project, which is a testimony to his kindness and selflessness.

Where did you grow up? Decatur, Illinois – I was born in Norfolk, VA when my father was in the Navy, (served in WWII). My parents left Norfolk approximately 3 months after I was born and moved back to their home, which was Decatur. My father's family were 11th-generation farmers in the Macon County area and my mother's family was located south of Macon County although still in Illinois. But my father and mother moved to the urban setting of small town Illinois, leaving their rural roots behind them. My father was a professional city firefighter, and my mother a homemaker. One of my most vivid memories is of my paternal grandfather, Isaac, who moved from being a farmer in Argenta to Decatur, and in retirement was the head groundskeeper for the Decatur professional farm baseball team, the Commodores. My family would attend games at Fan's Fields to regularly watch the team and see Grandpa.

What are your hobbies? I don't have many hobbies; I have been fortunate that my career has allowed me to follow my interests. I am intrigued in an incredible number of subjects; biology, zoology, anthropology, and mysteries, which have been the essential thread throughout my life. In addition, I study and collect art, architecture, tiki artifacts, souvenir buildings, art-deco, zoo albums, and animals, and I currently have 2 dogs, but I've had cats, and reptiles too. When I was younger, I thought I was going to grow up and be a zookeeper, and thus had a backyard zoo of several species. I enjoy science fiction movies although I am not interested in science fiction books and literature if that makes sense. In the past, I was a follower of Charles Fort's work.

Who are your favorite musicians or bands? I am a child of the 60s, I appreciated my parents' country and classical music tastes. My mom was a devotee of Elvis, so I listened to him at times as well. My brother Jerry would imitate Elvis, wear his hair and dress like him while imitating his voice. I enjoy the Beatles, Jimi Hendrix, Buffalo Springfield, Bee Gees, and Steve Goodman. I also enjoy bluegrass music. I would say like many of my interests I am diverse in my music choices. Also, I enjoy ethnomusicology and Tiki-related music.

What is the favorite place you have visited? Loch Ness, as a cryptozoologist I had always wanted to visit. In 1999, I was invited to keynote the first *International Cryptozoologist Symposium*. I took my two sons Malcom and Caleb, and while there we did participate in a surface expedition. I was supposed to join Dan Scott Taylor on a below-surface expedition in his 2-man submarine, although days prior he shared with the media that he was planning on harpooning the Loch Ness Monster when he found the creature. When the Scottish authorities heard what he was planning, they revoked his license for the expedition, and the trip never happened. After I returned home from the symposium, I spoke to Dan Scott

Taylor and he mentioned that if they would have taken the expedition, I would have needed an umbrella because they could not get the hatch of the submarine to close correctly. It was a blessing as I just thought of my 2 sons that may have been waiting for me to return and that we may have never made it back to the surface.

If you could take one person past or present on an expedition with you, who would it be? Ivan T. Sanderson, he was a friend and would have been fun. He was very experienced in the jungles of the world.

When and why did you get involved in this? In March 1960, there was a series of tv movies that were aired on Friday nights and broadcast again the following morning. One was called *Half-Human*, directed by Ishiro Honda (also known for *Godzilla, Rodan,* and others). *Half-Human* was about Abominable Snowmen in the mountains of Asia. I was fascinated by these things called Yeti or Snowmen. After watching this, I remember I went to school the following week and asked my teachers, "What is this business about Yetis or Snowmen?". I received 3 answers from my teachers; "they don't exist", "leave me alone", and "get back to your schoolwork". These may have been discouraging answers although they did not dissuade me.

I decided the Decatur Library was an excellent choice to research my interests. So, after I received the initial response from my teachers, I asked the librarians about the subject. Back then the term "Crypto-zoology" was not used, so they sent me to a section in the library called "Romantic Zoology." I asked many questions and was always looking for books. Through this, I was very well known and became friendly with the reference librarians. I had already been reading books by Roy Chapman Andrews, Raymond Ditmars and others regarding animals and animal mysteries. The books I had found in the library at that time were written by authors who covered general zoology, although, within those books, like one written by Willy Ley, there is a chapter dedicated to the Abominable Snowmen. There were other similar ones as well, for instance, by Herbert Wendt. I finally started going to barber shops where they would sell me the used magazines, *Argosy, Field and Stream, True,* and within those magazines were the 1950s and 1960s articles on Bigfoot. I noticed Ivan T. Sanderson was writing articles on the topic in *True* magazine.

If you have had an experience(s), what was the most compelling one? Many people get started and deeply into Bigfoot research and cryptozoology after having field experience. I of course had the popular cultural experiences which started my interest. Through the years, I had found tracks in Southern Illinois, I then started studying the creatures known as Skunk Apes reported from Illinois to Florida.

There was one incident in 1969, I was working in a mental hospital when myself and my co-workers were ride-sharing and on our way, one night after our shift was finished. We were traveling in a Volkswagen Karmann Ghia and I was in the back seat sitting on the middle hump. As we were traveling, we saw a giant full-sized Black Panther cross the road. I tried to convince everyone to turn around so we could investigate although they responded, "Are you kidding, we would rather go home and have a beer". Most people did not have the interest in the subject like I did. An interesting fact I share with patrons of the museum is that the most numerous sightings of any cryptid in the United States are of Black Panther, not Bigfoot.

What was the most compelling witness account that you heard first-hand? My father would get me introduced to game wardens, so at the age of 13, I was already investigating reports of Black Panthers and Bigfoot. I chose to attend college in Southern Illinois, so I could be close to where reports were taken. I would skip classes so I could investigate sightings of cryptids in Illinois and surrounding states. I was so curious about the subject; I couldn't stand still.

One case that always sticks out, was the Dover Demon in Dover, Massachusetts, in 1977. I was the first one there to investigate, as I was working in a school close by, interning for my graduate degree and working night shifts as well. At times, I would go across the river to the Dover Country Store to see what type of cryptozoological or strange antiques they had. I was in the store at the time of the Dover Demon sighting. The clerk's name at the store was Melody Fryer; she had posted a drawing of a creature that a witness claimed to see in April of 1977. I tracked down the witness along with 3 other eyewitnesses and created a case file. I decided to call this creature the "Dover Demon," it was a little orange thing that moved along a wall. I subsequently obtained help from three other experienced investigators. To re-verify the accounts, we all re-interviewed everyone. We completed a deep investigation. One of the investigators was the assistant planetarium director at the Boston Science Museum, Walt Webb. We created a high-level detailed report and concluded that we did not know what this creature was. I thought this was a remarkable and interesting story. Over 40 years later, I was shocked to discover there were Japanese toy replicas made of the Dover Demon; it is one of those cryptids that continues to live on.

What do you feel Bigfoot are and why? I always like to start out by saying we must be honest; we really don't know what Bigfoot are and we will only be certain once we have verified evidence. I know some scientists and Bigfoot researchers feel they are *Gigantopithecus*, and apparently based on the fossil evidence of a possible giant ape. I have always had a similar opinion of hominology theorist, Gordon Strasenburgh. In the 1970s he theorized that Bigfoot was a *Paranthropus*. Sasquatch/Bigfoot certainly appear to be more closely related to *Paranthropus*. I feel, the Yeti are most likely Gigantopithecus, and the Skunk Ape of the Southern United States are more chimpanzee and possibly a completely distinct species. They may have been transported here in the past by slave ships.

If Bigfoot is a large primate, what environmental conditions would it need to survive? Firstly, a "psychological" Bigfoot may exist, everybody thinks that every state has one, which most likely is not the case. The best concentration of actual physical Bigfoot are in the Pacific Northwest up through Southern Alaska. I don't think there are many of those around, maybe 4000 or less in that region. Their habitats are the border forests between the northern United States and southern Canada. These areas house a string of evergreen trees and hardwoods and would be viable environments. Although in areas like Maine for instance, if there are Bigfoot, it may be an interloper from Quebec or a migratory species. Out East, they used to be called Windego (sometimes known as Wendigo, Witiko, and Witiku), I feel these have mostly vanished and possibly become extinct. Furthermore, ancestry memories and folklore may still exist which keep them alive. Several sightings that are reported, may be misidentifications, sometimes people make visual mistakes while in the woods.

How would you explain the elusiveness of Bigfoot? I agree with the late Grover Krantz and others, for every 10,000 black bears there is most likely a single Bigfoot. Their population is so low, it is no surprise that there are no more sightings. Most that are seen crossing the roads are tall thin males which could mean they are adolescents and being pushed out of their family group. Sightings of more than one Bigfoot, typically come from the reports of people who are hiking deep in the woods, where the family groups of them are. There may not be many family groups in existence based on reports and historical accounts. They stay in small groups not tribes and seem to not want to be around humans. Another reason there may be fewer reports in the East is that it seems they do not like dogs and with the human population density in the East there are more pet owners with dogs.

What are your current research goals and how do you go about them? I have my blog, *CryptoZooNews*, and many people also look to me to write obituaries for the Bigfoot researchers and cryptozoologists, I started writing these over 25 years ago to ensure that we paid tribute to the researchers we have lost. In my blog, I also share the top cryptozoology stories and books. I started sharing these lists in 1999 and continue today.

At the museum, we have traveling exhibits and continue to have more coming in on a regular basis. I am currently looking at property to expand the museum to a new location, we are running out of space, so I want to ensure that have adequate room to preserve the history of cryptozoology. For the project, Doug Hajicek has designed the model for the future museum, and we are going to start fundraising soon. This will allow us to double our space.

In addition, I am writing 5 books; one on Merbeings with the late Mark A. Hall, one on general cryptozoology, and a few others focused on other specific areas of cryptozoology. I am bound to keep on going [laughing] and continue to also present and attend conferences throughout the country. I have been told a few times that I am the *"Old Guard"* in this field, I was personal friends with many of the researchers that are not with us anymore (Green, Bindernagel, Titmus, and others) and now I feel fortunate to bridge the gap and work with a new generation of researchers. I was just on the cover of *Maine Seniors* magazine, which is incredible, I never see myself that old [laughing].

Authors note: after this interview was conducted, Loren opened the second International Cryptozoology Museum in Bangor, Maine.

What do you feel is your strongest skill/talent that helps you in this field? My strongest suit is I am very out-spoken, very concrete, and not wavering and yet I am anthropologically and zoologically based in my theories.

I know that I am a radical, and immensely proud to be skeptically open-minded. Furthermore, I am aware that there are many misidentifications and hoaxers. When I lead an expedition, I am always dubious about adding individuals on both ends of the spectrum in terms of the true believers (those who think every sound they hear is a Bigfoot), and the debunker skeptics (who can't be open-minded enough to let any added information in).

International Cryptozoology Museum Portland, Maine. Wood sculpture by artist Snuff Destefano. Photo courtesy of the author.

In addition, I am also outspoken about the idea that there may be more than one type of Bigfoot. I feel the same could be said for Lake Monsters and Sea Serpents. One of my early challenges in the 1980s was that Grover Krantz and Jeff Meldrum pushed back a little on this theory, they felt in the field of Bigfoot, there was most likely only one type to be discovered. They said I was a problem because I felt that there was potentially a variety of anthropoids and hominoids out there. I would continue to say, "yes I do". Among Native Americans, there may be more than 400 names for Bigfoot, partially due to 400 plus tribal groups. Looking at the biosphere, the habitat, and we cannot assume this is all caused by one type of hominid or hominoid.

I do feel anthropology has caught up with cryptozoology, we are now seeing that paleoanthropologists are saying that there were 6 different ancient hominids existing at the same time.

What do you feel is your biggest challenge that hinders your goals? Much of the fringe material is draining away money and funding. We need funding to preserve the research materials, conduct DNA tests, and support expeditions. In addition, if individuals or groups are concentrating their time and money on non-research-oriented goals (social media, merchandise, and far-out distractions), it is not science and subsequently not pushing the field forward. Approximately 25 years ago I was interviewed by the *Chronicle of Higher Education*. They asked me if I thought cryptozoology would become a degreed program. I said "Are you kidding? There are not many courses in cryptozoology now and certainly not enough for a department to give credentials for a degreed program in it. We must be realistic and become more serious to move this field along." I still consider this is the case, academia's stance on the subject hasn't changed much.

What is your elevator pitch for open-minded skeptics? When it comes to the subject, I am not evangelical, I think being evangelical is the wrong approach to convince skeptics to be open-minded. I am not out to sell the science either, if people are curious and interested, they will go in that direction. I feel that people who want to debunk are not critical thinkers. Cryptozoology will really thrive most in a critical thinking environment. When I have patrons in the museum that have some type of interest in the subject, I show them our display, *The Classic Animals of Discovery.* This display contains the animals that were past cryptids and now are identified newly verified species (Okapi, Pygmy Hippopotamus, Giant Squid, Mountain Gorilla, Saola, and others). We also know that once a cryptid is discovered and verified, cryptozoology loses that animal to Zoology. We are discovering new animals all the time.

What suggestions do you have for researchers who want to get involved in this field or further educate themselves? I suggest reading books from authors from around the world. We engage the patrons who enter our bookstore so we can direct them to the books that match their interests and to the authorities they can engage. Our goals are to encourage education.

How do you feel mainstream scientists will publicly accept the existence of Bigfoot? Many interested researchers and scientists may already be part of the so-called "Cryptozoology Academy," but they are now teaching in a manner they weren't years ago. I remember when I went to college, my anthropology professor had a challenging time with my paper on *The Surviving Neanderthal,* he gave me a B- because he thought the subject was too radical (although I included 60 references instead of the required 3). Some professors act like they are skeptics publicly although privately they are interested and take part in expeditions and the associated studies. Furthermore, some skeptics are starting to author books on cryptozoology so there is a grain of hope that if some of them are presented with physical evidence, they will take the time to look at it. The DNA studies by the late Brian Sykes certainly went in that direction.

Some think technologies will provide the final proof of Bigfoot, I don't think so. In some cases, technology and social media have been to the detriment of cryptozoology. I do think environmental DNA and other types of DNA testing may be a ray of hope for discovering new animals without killing them. Mostly I feel it will happen in the oceans and not on land with Bigfoot. I do think that the Orang Pendek in Sumatra will be the first discovery of a small Bigfoot-type animal.

Once a discovery is made in anthropology it becomes part of the mainstream and soon does not become revolutionary anymore. For example, The Hobbits or *Homo floresiensis* were a remarkable fossil record discovery in Indonesia and so were the *Homo luzonensis*, the small fossil people of the Philippines. Both are now taken for granted. Most of the islands in the Pacific have local living reports of Menehune, Ebu gogo, and other little people and now will be explained quickly in anthropology as future found groups of fossil humans. Thus, Cryptozoology becomes tomorrow's Science and Anthropology.

How to follow Loren Coleman:
Twitter: @CryptoLoren
Cryptozoologymuseum.com
Cryptozoonews.com

DARYL COLYER

Daryl Colyer is a former Airborne Intel Operator in the United States Air Force; he flew dozens of missions in very remote and far-flung places around the globe. He was trained by the Air Force in combat wilderness survival. He has a BA degree in History and is currently in graduate school earning his MA in History - War & Violence. His civilian career post-military has been in banking. He is also a lifelong outdoorsman, hunter, and adventurer.

Daryl is well-spoken, meticulous, and methodical; I was fascinated with our conversations and his knowledge of the subject. His background adds integrity to the research he participates in and credibility to his past firsthand experiences.

Daryl currently serves on the Board of Directors for the North American Wood Ape Conservancy (NAWAC) and has been with the group since shortly after its inception two decades ago. The NAWAC is stringently dedicated to conducting scientific research, it does not seek public notoriety, and has an intense, no-nonsense focus on definitively answering the wood ape, or sasquatch, question. Daryl continues to work within the team structure of the organization toward legitimizing the subject and removing any stigmas attached to it.

Where did you grow up? I am originally from Northeast Texas. I was born in a little hospital in Atlanta, Texas, which is not far from the infamous area of Boggy Creek, or Mercer Bayou, in Miller County, Arkansas. In fact, we lived just outside Bloomburg, Texas, only a few miles from Fouke, Arkansas. When I was young, I routinely heard stories about the co-called Fouke monster.

What are your hobbies? I'm currently in grad school. My major is history. My focus has been largely on medieval history, war, and violence. My professors are strongly encouraging me to pursue my PhD. If I do that, I will focus on the Cold War, of which I have strong firsthand knowledge and experience. I am also into fitness, the outdoors, and the St. Louis Cardinals baseball. I am a Christian and I am active in church as part of a music team. I play acoustic guitar and sing. I also cherish spending time with my wife on our four-acre rural Texas property.

What's your favorite musical genre or band? I love all kinds of music. When I study or meditate, I invariably listen to Post-Baroque Renaissance Music (Johann Sebastian Bach, Antonio Vivaldi, Arcangelo Corelli, Alessandro Scarlatti, Henry Purcell, Juan Gutierrez de Padilla, those types of timeless composers). The music from that era, the early modern period, seventeenth to the mid-eighteenth century, had such a vast degree of richness that is missing from music that was subsequent to it. Now, if I'm working out, I'll listen to rock pop from the late nineties to early 2000s (Michelle Branch, Coldplay, the Frey, Hoobastank, Death Cab For Cutie, Mat Kearney, Avril Lavigne, Snow Patrol). I love some 70s – 80s rock, too:

Eagles, Boston, Genesis, Mike and the Mechanics, Crowded House, Don Henley, Toto, and so forth. I also deeply love country, bluegrass, and Christian music.

What is something that most people do not know about you? Well, I worked closely with Mercury-Polygram Records in the 1990s after getting out of the military on a record deal, but it just never fully developed.

What was an embarrassing or funny memory in your past? I threw up in my trainer's helmet on my first flight, somewhere in some remote hostile place with enemy fighters buzzing all about.

If you could take one person past or present on an expedition with you, who would it be? My wife.

When and why did you get involved in the subject of Bigfoot? Well, like I said, I grew up hearing many stories and when I became of age, joined the USAF Intelligence community, I really started to question those sorts of things.

One time, we were waiting to launch a mission. Our plane had been grounded until some work could be done on it. I was sitting there with the aircrew and just started discussing diverse topics and this subject came up among many. We got deep into the topic and I sort of made a vow to some of the Silent Warriors. I told them that if the thing is real, that someone should be able to dedicate weeks to finding these things if they existed. I told them that once I was out of the Air Force that I would go somewhere where people claimed to see them and I would get to the bottom of the mystery. I had graduated from SERE training, I had, all my life, fished, hunted, camped, and spent all kinds of time out in the woods and in the outdoors when I was growing up. I thought, "I'm equipped for it, I know how to operate in a wilderness environment, so that's what I'll do."

Fast forward, a decade goes by and I forgot about all that. In the early 2000s, we were taking a trip west and went through New Mexico, which is just a vast state and doesn't have a whole lot of people. We went through this big wilderness area and my daughter asked me out of the blue, "Daddy, do you think Bigfoot lives here?" It's a simple question, but it reminded me of what I had told my fellow airmen.

When we got back home, I started popping around on the internet and I found a report from this dude named Alton Higgins. I originally thought he and his name were fake. I thought, if he's a real dude, then I can find him. I tracked him down at a university in Oklahoma. I called him on the phone without prior notice and we talked for a couple of hours. Soon after that, I drove up to meet him. We struck up a friendship and I was soon joining him out in a highly promising wilderness area. Ever since then, I've been neck-deep in this stuff.

Can you share your first experience with a Bigfoot?

Daryl Coyler Audio

After your visual encounter, how did you process it? It took me several days to process it and come to terms with it. You just really try to talk yourself out of it. Most everybody I know has done this same thing. Even though I was already involved in trying to solve the mystery, I was not at all a "believer." I felt I couldn't just dismiss it out of hand; I had to thoroughly investigate it, apply critical analysis to it, and go wherever the investigation led me. That's what I did, and at the time when I saw that thing that day, I was not at all convinced that we were dealing with a legitimate species. I still had serious reservations about it. If I had to quantify it, I was probably 50/50, real versus myth, at that time. I kept thinking there was no way this thing could actually remain hidden and remain undiscovered thus far. I just couldn't wrap my head around it. After that initial visual down in Southeast Texas near the Trinity River in Liberty County, I was stunned and had a hard time admitting to myself what I had seen. But in the end, there was no denying that I saw what I saw. I had the visual, heard it running off–it sounded like a person running off in the woods–and it had a raw, unique, distinct odor. The odor was kindred to other smells–like known great apes, or a "zoo" smell–but at the same time, it was a unique smell all itself.

What do you feel Bigfoot/Sasquatch are? Everything I have observed and documented supports the extant unrecognized hominoid hypothesis.

If Bigfoot is a large primate, what environmental conditions would it need to survive? Expansive old-growth forests with little or extremely limited human encroachment.

How would you explain the elusiveness of Bigfoot? They are rare, extremely intelligent, very ambulatory, and they have extreme physical capabilities. They are highly wary, extremely sensitive to changes in the environment, and reside in remote pockets of North America with very arduous terrain.

Do you feel since the 1960s that we have gotten any closer to proving the existence of Bigfoot and why? Yes, I do. In 2015, the NAWAC released its *Ouachita Project Monograph*, a summary of our findings up to that time in the Ouachita Mountain Ecoregion, a place we call "Area X."

What type of research are you currently involved in? We remain dedicated to collecting irrefutable empirical evidence that the species exists. We are engaged in long-term field studies with dedicated field teams in a singular environment where we have documented encounters with the species repeatedly over a long period of time.

How did the NAWAC get started? It started as the Texas Bigfoot Research Center (TRBC). We changed the name to Texas Bigfoot

Research Conservancy because we believed it was our mission to not only validate the species, but they would also work to preserve critical habitat. Then after several years, we wanted to expand our membership and our scope; thus, we became the North American Wood Ape Conservancy. We made the decision to choose the term "wood ape" because we believe the term best fits the animal(s) we've encountered, but also because the words bigfoot and sasquatch carry a stigma with them. Based on our observations, and our own experiences in *Area X* and other places, we hold to the working hypothesis that this is some sort of undiscovered ape, possibly an extant ape that has been here for a long time, or it's a species with which no one is familiar, a totally novel species.

What scientific approaches do the NAWAC take in research techniques? The scientific method is about observation, that's where it begins. One begins with an observation and from that emerges some sort of hypothesis. Next, you look to prove or disprove that hypothesis, which is essentially what we are in the process of doing. We are trying to validate the extant hominoid hypothesis, that there is an unknown hominoid species behind the reported visual encounters. We have not been as focused on obtaining visual evidence as we should have been in the past, but we are changing that with a new focus toward obtaining a portfolio of clear, high-quality photos/film. We believe if we are able to do that, the possibilities will be exponential. In turn, those prospects likely will lead to the definitive empirical evidence that the conventions of science require.

How to follow Daryl Coyler:
www.woodape.org

SHANE CORSON

Born in Scotland, Shane Corson moved to Southern California when he was a teenager and later to Oregon after he got married. He was fascinated with the subject at an early age and later started to investigate and interview witnesses. In 2011, things changed for Shane, he had his own riveting sighting which further fueled his passion to solve the mystery. It eventually led him to become a team member of the Olympic Project which was founded by Richard Germeau and longtime Bigfoot researcher Derek Randles. This group of resolute individuals has uncovered compelling evidence of the possible existence of Bigfoot including visual sightings, tracks, audio and 2 potential nesting areas.

Shane's passion for the subject is contagious, he is humble and extremely gracious to the researchers that have influenced him. He is one of the hosts of *Monster X Radio*, is a member of Project Zoobook, and has been featured on multiple podcasts and media outlets over the years. He and the Olympic Project team were recently featured on Small Town Monsters, *On the Trail of Bigfoot.*

Shane continues to be boots on the ground, he gets out in the woods extensively and is a proud representative of the Olympic Project Team. The impactful research and discoveries he and the team have made have opened the eyes of scientists that previously would not have taken evidence related to Bigfoot seriously.

Where did you grow up? I grew up in Scotland on an island in the Outer Hebrides, off the west coast of Scotland called Islay. It's known for its whisky, fishing, and an amazingly rich history. It's just a beautiful little island with a couple of towns here and there.

What are your hobbies? I have been an avid fisherman my entire life. I enjoy hiking, camping, and love hunting. Having played a lot of soccer in Scotland and the U.S., soccer is a huge part of my life, as is the outdoors. In addition, I study paleontology and love anything related to dinosaurs.

What type of music do you enjoy? I love rock and roll, country, Scottish folk music, classical, and anything bagpipe related. I have been a huge Pavoratti fan my entire life. In my younger days, I was into the dance scene but I still like to get out and boogie. Nowadays it's smooth sailing with country and classic rock.

Why did you get involved in the subject of Bigfoot? Growing up in Scotland I was really interested in paleontology and animals in general including cryptids, like the Loch Ness Monster. I had heard of the Yeti and cryptids from around the world, including Sasquatch or Bigfoot. When we moved to the States in 93, I was in Southern California. I read everything I could get my hands on, every documentary, every everything cryptid related. Being in California, it was a given that I would dive into the Sasquatch phenomena, especially knowing that the Patterson Gilman film was filmed in Northern California.

In 2006, I eventually met my wife, and we moved up to Oregon in 2008. I wasn't too far away from the coast or the mountains, specifically Mt. Hood. So, I started getting out more, camping, exploring, fishing, hunting, hiking, and everything else I love to do. I was kind of a solo runner back then when it came to investigating. I'd look for trace evidence, tracks, impressions, everything I could get my hands or my eyes on. I never really came across anything of interest until 2011 in Mt. Hood, when I had an encounter with these things and had a sighting which solidified the existence of Sasquatch for me. Eventually, though, I concluded that this is not a one-man endeavor and that's when I started reaching out to people like Cliff Barackman and Derek Randles specifically. That launched me into where I'm at now with the Olympic Project.

Can you talk about that experience you had in 2011? I was on a remote backpacking trip with two friends in the Mt. Hood National Wilderness, in a high mountain lake area. Neither of my two friends were really into the whole sasquatch phenomena. On the first day, we hiked in the early morning and set up a base camp near a lake. After setting up camp, we decided to go and try to find one of the other seven lakes where we planned to fish. We ended up getting lost and it took us the entire day to finally get back to base camp. One of my friends, Ian, was afraid of bears so we built a huge fire and got to bed around 11:30pm. We had three separate tents, kind of in a triangle formation set up a few feet apart next to this lake.

Around two in the morning, I heard what sounded like two rocks being clanked together. Just prior, it was dead quiet outside, you could hear a pin drop. The clanking noise got closer and closer. My other buddy Mitch whispers over to me, “Hey, do you hear that?” We tried to rule out any native animal sounds. Whatever we were hearing, was stomping around in the woods, and seemed to be circling our camp. It sounded large and bipedal, but it was hard to know for sure. That went on for a minute or two and then it stopped, and then the clanking sound started again eventually getting further away and fading before we fell asleep.

The next night at camp, we hear the same thing, clanking and stomping at approximately the same time. This time though it got closer and seemed more purposeful, it was louder and we heard fallen branches being broken by this creature while it walked. Like the night prior it was definitely quiet again outside except for the stomping and clanking. Suddenly, we hear five powerful thuds on a tree, or power knocks is what people like to call them. It echoed through the valley and we could feel the vibration underneath our tents. We then hear what seemed like a rock being thrown through the trees landing next to Mitch's tent. At this point, Ian was freaking out, so I decided I was going to unzip my tent to console him and calm him down. As I'm peering out, I see some movement out of the corner of my eye. There was a large Douglas fir tree in front of us about 30 or 40 feet away. As I'm focusing, I see something very large swaying back and forth. I can see a hand, fingers, an arm, a shoulder, and a head. It was popping out from behind the tree, swaying, I could see the shoulder and head every couple of sways. I'm sitting there and I'm petrified because I'm looking at this thing and it's massive and I'm thinking, what's going to happen? Then it removes its hand from the tree, drops it down to its side, turns around and walks up the trail and that was it. I sat there for a while and never got out of my tent.

As soon as daylight hit, we all collectively, without even speaking, packed up our stuff and headed out. We were supposed to spend another night or two out there, we had made plans to go to another lake, but everybody was pretty freaked out. We barely spoke the entire trip home; weeks passed before we discussed it with each other.

What is the Olympic Project? The Olympic Project is an organization of dedicated and like-minded individuals who are committed to documenting the existence of sasquatch through science and education. We are a data collection-driven organization first and foremost. We have members from all social classes; academics. archaeologists, biologists, primatologists, hikers, hunters and outdoorsmen and outdoorswomen. Most of the individuals in the

Olympic Project have either had a sighting /encounter or the evidence that they've come across has led them to delve into this phenomenon and take it seriously. We all carry a skeptical eye and try to thoroughly vet the findings and the data we collect, utilizing Occam's razor. We look for patterns of predictability in the why, the who, the what and the where. The Olympic Project is not necessarily out to prove the existence of Sasquatch, but in case that day of verification comes, if sasquatch is ever proven to exist, we will have a plethora of data readily available for academic and public consumption. The project is made up of just a good group of people, we get along great, we're all very ambitious, and have the same ideas, goals, and passions, and most of all we have fun.

Olympic Project Study

Do you think the nesting sites are the most compelling evidence you have found out in the Olympic area? Hands down, 110%. It is not just the most compelling and interesting thing to me; I feel it's one of the most fascinating and convincing finds in the history of the subject. There have been other nests found and discovered over time, but never this many in one area and never studied to this length. Usually, there's a picture taken, a story told, nothing's collected, and that's it.

Can you describe the nests? A few of the nests were originally discovered by a timber surveyor back in May of 2015. This individual was amazed and taken back by the look and the sheer size of the ground nests. In his 27-plus years of being in the timber industry in remote places, he had never seen anything like them. This led this individual to reach out and invite Derek Randles of the Olympic Project to take a look at his discovery. Derek Randles and James Million of the OP were joined by the owner of the timber company, two DNR (Department of Natural Resources) individuals, and the original surveyor, to hike out to the location. They were all amazed by the ground nests and had never seen anything like them in all of their years of being outdoors. The OP was then given permission to study the nests and the surrounding area.

To date, there have been 24 nests discovered. All but two of the nests were constructed on the ground. Two nests (or practice nests as we like to call them) were constructed in actual, huckleberry bushes, three feet off the ground and a mirror image of the ground nests. The nests were discovered on multiple fingers, in clusters coming off of a large ridgeline that sits above a seasonal salmon creek, 2.5 miles behind a locked timber company gate and way off-trail. All of the nests are constructed and built from nothing but evergreen huckleberry bows. The vast majority of the nests are bathtub in shape, ranging in size from about 4ft to 8.5 ft in length, with larger branches (1 inch to 2.5 inches) being on the bottom as a base layer and smaller branches (less than an inch) on the top that formed a mattress. Some of the huckleberry branches were pushed into the ground in a stake-like fashion and the huckleberry bows were weaved or formulated around the ground stakes as a sort of bed frame.

We surmise based on in-field observation and data collection, that these nests are being made in late February to early March and possibly every 4-6 years. Hair has been collected from these nests that does not match any known animal in the area and visually looks to be primate and lacking a medulla. We have also collected hand impressions, and multiple foot impressions, and have recorded a ton of very compelling audio from this area. We believe this evidence all points to unknown behavior from an unrecognized species.

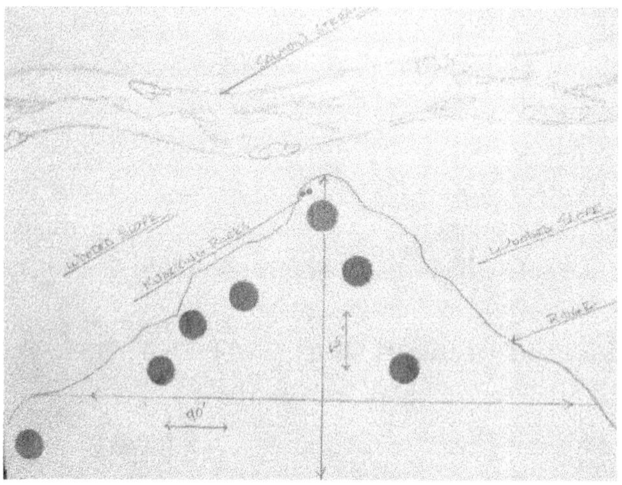

Nesting Area – Photo courtesy of Shane Corson

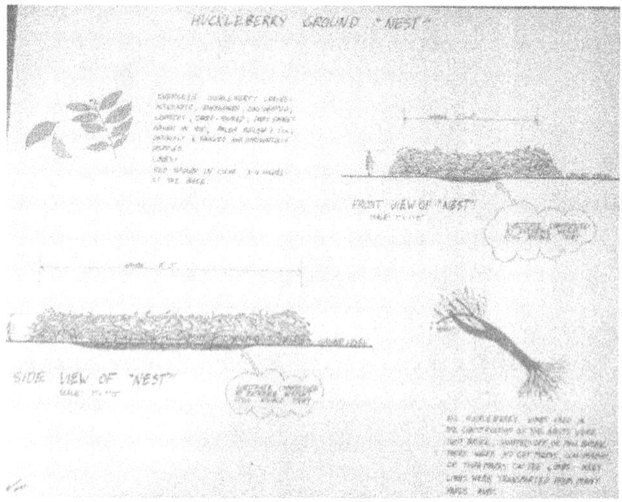

Nest Diagrams – Photo courtesy of Shane Corson

What are your suggestions for new researchers? First, have fun with it. If you are not enjoying yourself then don't do it. Have a sense of humor, take it seriously but enjoy what you do. If you decide to work with other individuals, make sure you have the same goals and ideas. Try to absorb as much information from past and current researchers and learn the scientific method. I'm a student and an observer. I encourage people to learn as much about what's been done in the past and those that came before them. Be skeptical and thorough, and never jump to conclusions. Know how to spot a hoax. Research those types of things before you really start getting in the field.

Once you find a location and prepare to investigate it, study the known animals in that specific area. Make sure that you take dubious notes, and document as much as you can. Even if there's no historical sasquatch reports in that area, although it does have an abundance of wildlife and food sources, ask yourself, "If you were a sasquatch, would you be in that area?". Does the area have an abundance of wildlife, water and food sources, and concealment? Don't ambulance chase, I would focus on data collection, know your research area, practice your outdoor and survival skills, be prepared for anything. Always tell somebody where you're going when you're coming back, and again, most of all, have fun with this.

Tread carefully on social media if you can or avoid it altogether because it is a dog-eat-dog world, and sasquatch will not be proven to exist on any of these platforms. Some investigators and researchers live to attack and belittle each other. There is that element out there, and some people do not get along. Your time is better spent in the woods or reading something about known wildlife or about the Sasquatch phenomenon rather than engaging in an argument online about something that's not even proven to exist yet. Truth is, you don't owe anybody anything. Remember, do it for yourself, have fun at it, and good luck.

How to follow Shane Corson:
Olympicproject.com
Monster X Radio
Facebook: Shane Corson
Instagram: shano_corson

MARC DEWERTH

I have mentioned the Ohio Bigfoot Conference (OBC) a few times in the book. Most feel it set the standard for several of the conferences we have today on the subject. The groundwork for conferences in the state of Ohio began with Don Keating dating back to 1989, the first one held in Newcomerstown. He eventually moved his conference to the Salt Fork Lodge in Cambridge, Ohio. Salt Fork State Park has a history of Bigfoot encounters. In 2012, Marc DeWerth took the helm, changing the name and creating the OBC. The event continues to be held at Salt Fork Lodge and attracts enthusiasts and researchers from across the country.

Marc has been researching the Ohio Bigfoot phenomenon for over 30 years, is the Ohio curator for the Bigfoot Field Researchers Organization (BFRO) and has interviewed over 300 alleged witnesses. I first met Marc at the OBC and have gotten to know him in the years since. Self-admittedly he may be testy with what many call the *Bigfoot Community* and the lack of camaraderie and unity, although he does a respectable job organizing and running his event and bringing people together. He continues to have the top researchers in attendance and is dedicated to educating the public on the historical accounts of Bigfoot in Ohio and around the country. Marc and the OBC have provided me the opportunity to meet many of the researchers in this book and other individuals who have become friends.

Who are your favorite musicians or type of music? My favorite band would be AC/DC, my favorite musician is Rick Emmett, who is of the band Triumph. He is a great guitar player with a lot of diverse playing styles. I also love all sorts of diverse types of rock and roll.

If you could take one person past or present on an expedition with you, who would it be and why? René Dahinden. He was absolutely the most dedicated field investigator of his time, hiking 100s of miles looking for Sasquatch (and never seeing one). Very funny, even more nasty, but brutally honest. I was honored to have known him.

When did you get involved in Bigfoot research? When I was a little boy in the late seventies, and early 80s, every Friday night my grandpa and I would always watch *In Search Of* the show with Leonard Nemoy. After seeing the episode of Sasquatch in North America, I was extremely interested in Bigfoot, and I asked my grandpa if Bigfoot was in Ohio, and he said yes. I was blown away that my very smart grandfather would think that. He said, "I've seen stories in the newspapers about it".

Then, I think I was in fourth grade at Hilliard Elementary. I asked the librarian if they had any books on Indian folklore because I was afraid to say the "Bigfoot" word. They put me over into a section where I noticed the Loch Ness Monster book, so I knew I was in the right place. Then I saw the word Sasquatch written on a green mylar dust jacket and it was John Green's book, *Sasquatch Apes Among Us.* I remember looking on the back of the dust jacket where they had a map of the North American continent, and I saw Ohio. It showed 17 or 18 sightings and I checked that book out right away and started reading. That was literally the first book I couldn't put down, I was hooked.

If they exist, what do you feel Bigfoot/Sasquatch are and why? They are an unknown form of hominid that has survived all these years virtually undetected by everyone except the local Indigenous people. Over many years, facts became stories, stories became legends, and legends became sightings.

How would you explain the elusiveness of Bigfoot? They are so tuned into the environment. Their smell and eyesight must be incredible, as they detect our presence long before we have any clue of them being around. They're so active at night putting them at a huge advantage to our reliance on technology ways. True Rambo's of the forest.

What excites you about the subject? What keeps me going is the sheer interest in what this thing really is. Secondly, the relationships I have built with people and now that have become my friends. Then, of course, I like being out in the woods all the time. So, I figure while I am out there, I might as well look right?

What researchers past or present have influenced your work? John Green, Rene Dahinden, Daniel Perez, Don Keating, Ron Schaffner, and Dr. John Bindernagel.

What type of research are you currently involved in? Multiple properties with a combination of LDRs (Long duration recorders), and game cameras.

Are witnesses reliable? I've conducted onsite interviews with over 300 witnesses and taken thousands of calls over the years, you weed many of them quickly. Typically, 9 out of 10 are bogus, misidentifications, or where something just isn't right about the story. That's what is unfortunate about today, it's so much easier to hoax with technology and social media. Plus, there's so much information out there on websites so people can see what Bigfoots are said to act like, walk like, and so forth. They only need to be good story writers. That's why you must really filter the information that's out there. The worst ones are the individuals who make their rounds on podcasts and other media while their encounters have previously been debunked. Yet they're still making their rounds like their encounter really happened. Then you see the same videos or pictures turn up that were all the time. They get debunked and keep regurgitating and all the new people just buy right into it because to them, it's just excitement.

Is there a compelling witness account that stands out to you? I interviewed a bow hunter in Adams County, Ohio. He was up in his tree stand near an area where he also set up a trail camera. He was getting pictures of these massive bucks that kept coming through this area. His stand was positioned in a prime location where animals travel toward this drawing area to get from one point to another. He heard something huffing and puffing and making all sorts of noise, and it sounded angry. He started to get concerned. He is sitting in his stand which was approximately 11 feet off the ground. Suddenly, he heard what sounded like a bowling ball coming up over a hill, and he knew right away it wasn't the deer.

He stated this thing peeked over the hill and headed towards him. He said it was the largest thing he's ever seen in his life and seemed angry because the hunter was up in the tree stand. He said the creature struck the tree and screamed. This guy literally lost his cookies right there from both ends. He says this thing was so big, it could have reached up and easily touched him. The creature swayed back and forth at the base of the tree two or three times then retreated from where it came from. It looked back once, yelled and then aggressively returned to the tree stand. The hunter said that the whole time, all he could do was just pray that he was going to live. He said finally the Bigfoot walked away and disappeared into the woods. It took the hunter 3 hours to even have the courage to come down.

When he did, he left his stand where it was and got the hell out of there.

What were your experiences with the early researchers? When I was young and single and worked in the airline industry, I could fly for free. So, every time I had an opportunity to get a week or two of vacation, I would take trips to the West Coast. I met Ray Crow of the Western Bigfoot Society, Larry Lund, René Dahinden and Peter Byrne. I went to B.C. multiple times, stayed with John Green, and drove all the way to Pullman to visit Grover. They were just a real, unique crew of people.

Staying with the Greens was special, John and his wife, June, were fantastic. John was an open book, all his files, all his evidence, he allowed me to ruffle through them and read everything. I will always remember having the opportunity to listen to the tape recordings of his interview with Albert Ostman.

René was hilarious, he lived in a trailer over by a shooting range. It seemed every other group of words were, "screw this, screw that, he hated this, he hated that". I'll tell you what though, for a boots-on-the-ground researcher, I don't think you'll ever find a Bigfoot investigator like him. He would go anywhere and everywhere looking for these things and all those years he spent in the remote wilderness of Northern California and British Columbia and never had a sighting.

Grover, he wanted to shoot one. Any time you went and saw Grover, you had to prepare for what I called, "his smoke barrage" because he smoked so much. He really wanted to prove that these creatures existed and he took grief from academia. Being a professor, he put his reputation on the line that these things are real. He was one of the first people, if not the first that found the dermatoglyphic ridges in the footprint casts.

Then of course there was Peter Byrne. Peter is a great guy, a gentleman. We would spend countless hours talking Bigfoot and drinking brandy and scotch. Sometimes it would be 5 or 6 in the morning by the time we got to bed. He is a nice man, once I lent him game cameras for his preserve in India. In return, unexpectedly, he sent me a bushman's hat as a gift, which was pretty cool. Peter will never stop researching, it just keeps him going.

How did they influence your research? They authored books, they all interviewed similar witnesses and they all had similar opinions, they all investigated the Patterson Gimlin film site. In the 70s and early 80s, Bigfoot was such a taboo subject you didn't want to admit you were into it. To find a place to go, a place to learn about it, those were the people you went to. They were the ones out in the woods, doing the work and interviewing hundreds of witnesses.

How did the Ohio Bigfoot Conference get started? Back in the mid-90s, I was doing research with Don Keating and he hosted an annual Bigfoot conference in Newcomerstown. We hadn't met in person yet and he had always invited me to his meetings. One day my girlfriend at the time and I decided to go. We walked into the meeting and we're all dressed up, all snazzy and everything, there's maybe 40 people in attendance. They all turned and looked at us all dressed up and I thought they were going to eat us. There was dead silence and stares. Finally, Don Keating said, "Are you Marc DeWerth?", and then we were welcome. There were 40 people or so and they were all talking about Bigfoot, it was amazing. Not only from Ohio but West Virginia, Indiana, Pennsylvania and I believe even Michigan. I heard all these stories, I was sitting there like a sponge thinking, "I can't believe it, there's people in Ohio and all these states looking for Bigfoot, and I never knew anything about it". I told Don Keating, "If you ever need any help, here's my number, call me, I'll come right down". He called me within a few days and the rest was history.

We wanted to expand the audience, so we started bringing in the famous Bigfooters to the conference and the event started growing. Eventually, it was moved to Salt Fork State Park and in time Don got out of it and asked me to take it over. I willingly did and just put my own touch on it. Don had done such an excellent job with it. I decided to expand the event because I wanted it to represent Bigfoot research from all over, the Pacific Northwest, Florida, Louisiana, Georgia, Massachusetts, etc...

I keep the size manageable. I like it that way, we have the huge flea market that goes all day Saturday, which you don't have to purchase tickets for. People can come to shop, meet people, it's a family event. That's what I always really wanted, an event where people could bring friends and family. Even if you can't afford tickets, you can still attend, meet the speakers, meet people, shake hands, shop, whatever you want to do.

Can you share a memorable time at the OBC? I think the most memorable moment at the Ohio Bigfoot Conference was when Bob Gimlin was introduced into the Ohio Bigfoot Hall of Fame. The crowd gave him this longest-standing ovation, and Bob just sat there with tears dripping down his face. The thing about Bob, he is the classiest man on the planet, and he will always take time to thank each and every person, he is just exemplary.

Charlie Raymond speaking at the Ohio Bigfoot Conference. Photo courtesy of Marc DeWerth.

What is your advice for new researchers? The thing I always tell people is to educate yourself on the basic ecology of your state. When

I say ecology: what areas of the state produce the most mass on the forest floor, which is food. For a bigfoot to exist in your state, they're likely preying on white-tailed deer and if and if they are, you need to have oak and hickory trees that produce lots of nuts or mast. So, you want to find areas like that, which produce the amount of food that something this large can survive. If you're researching in an area that's only pine forest with extraordinarily little nutrition, the likelihood of a Bigfoot being there is not good. For example, Ohio has many large white oak-hickory forests with tons of white-tailed deer, there aren't many natural predators. If you don't have an abundance of natural predators, what happens to the prey? They get bigger and meatier so Bigfoot in Ohio fills that niche in Ohio very well.

How to follow Marc DeWerth:
Facebook: Ohio Bigfoot Conference Salt Fork

DAVID ELLIS

David Ellis is on the brief list of individuals who have focused much of their research analyzing audio related to Bigfoot. Well respected by his peers, he is often the first they turn to with recorded audio suspected to be from the elusive creature. He graduated from the University of Washington in 1974 with a BA in Psychology. His interest in the subject began at an early age when his grandfather described seeing a five-foot-tall monkey on his farm near Vancouver, Washington. Later when he was 11 years old, David and a small group of friends had an encounter when something large was breaking limbs next to them while they were walking along a tree line. The animal then let out a 12-second scream that David describes as a cross between an elephant's trumpet and a lion's roar. It scared him and his friends to the point where they plunged to the ground and then sprinted out of the field. With the encouragement of his parents and a teacher, this ignited David's passion to study the possible existence of Bigfoot and the vocalization aspects of the creature.

I find the work that David is doing fascinating. Although I certainly struggled to grasp the science behind audio analysis, David had a knack for explaining it simply so I could understand. This type of research is often overlooked and it takes a special individual to dedicate countless hours to studying audio. He currently has over 8500 clips of potential Bigfoot audio. In addition to being a sound specialist, he is a talented track analyst and has interviewed hundreds of witnesses.

He has been part of the Olympic Project Team since 2010. David has been featured on television, documentaries, podcasts, and other media. There is no doubt that the audio data he continues to collect through his scientific approach is compelling, even to skeptics. Without David and the few individuals who focus on this type of analysis, researchers would not have the opportunity for their recordings to be studied professionally and help connect the dots between their audio and other collected evidence.

What was an embarrassing or funny memory while researching? Closing down the Olympic Project Bigfoot bar for the night. Many a funny moment has happened at the bar where only true stories are told. It is a bucket list for many investigators or expedition attendees.

If you could take one person past or present on an expedition, who would it be? Well, it would have to be two people. Recently I have become a grandfather to my grandson and granddaughter. I hope they will be properly introduced to the wonderment of the world outdoors. I would like to share my experiences with them.

If they exist, what do you feel Bigfoot are and why? The answer to that will be in the DNA recovered from a body. They are part of the evolutionary chain of hominids. The branches are becoming fuller with each passing year.

How would you explain the elusiveness of Bigfoot? They know us better than we know us. Our predictability is used to their advantage. They can put us at a disadvantage in the blink of an eye. While in a car at night I saw a black object on the side of the road. My mind said it was a stump but there was something odd about it. After passing we turned the car around and went back. It wasn't there but left a huge disruption in the brush.

What researchers past or present have influenced your work? Jeff Meldrum, John Bindernagle, John Andrews, Monongahela, Stan Courtney, Derek Randles, Paul Graves, Kristine Skehan, Jim Boudousquie, and Ann Bastin

What type of research are you currently involved in? Audio recording and analysis.

How did you get started with audio recording? When I started doing audio work, I was mentored by Stan Courtney and other players in the Bigfoot community. I started buying equipment and experimenting on my own. From the very beginning of my research into the Bigfoot world, I would always hone in on specific witnesses and stay with them for extended periods of time. I didn't want to *just* interview them, and move on to another witness. I felt that if there was something really going on, then that's where I wanted to be, I wanted to be at that location. That process yielded to me more information than just going out every time somebody had a new report and chasing it.

Did you find that these witnesses would have recurring encounters? Yes, and I have audio documentation to back that up. This last week has been nuts, it seems like everyone decided, "Okay, let's send stuff to David today". I've been inundated with audio and it's a good thing. It's feast or famine as usual. I just heard some interesting stuff today that I've never heard before. Is it Sasquatch related? I don't know, although some of the audio sounds like words and that's exciting. This is with a witness that I've been working with for at least six years.

How do you study the audio? A gentleman who keeps his real name discreet, most know him as Monongahela was in the BFRO at the same time as I was. We started communicating. I started sending him audio to get his opinion on it. He asked me at one point if I was analyzing my audio visually, and I had no idea what he was talking about. He then taught me about *spectra graphic analysis* and introduced me to software that could help analyze my audio more efficiently. Once I started to learn *what* a sound is, I could see it has a print. I could then identify what that source was. Now I do not have to listen to the audio, I can see it. Although I typically do listen and visually study them because I don't want to miss anything. I typically get to the point where I can just scroll through the screen and say, "Oh, that's interesting, oh, that is weird". It allows me to zero in on the abnormal versus the normal sounds.

Can you explain how you visually see the sounds? I'll have to send you one. *spectra graphic analysis* does three things. First, it gives the horizontal line, I'm not sure if that's the x or the y axis, but the horizontal line gives you time. Then the vertical axis gives you the hertz of resonance. Thirdly, where the color comes in is; the louder the sound, the more vibrant the color, and they shift from distinct colors. If it's in black and white, the darker the color, the louder the sound is. We can see all sorts of nuances on a spectrogram that you don't get from a waveform analysis, which is amplitude, it just shows you the spikes. Waveform will only show how loud the audio is, it doesn't give you any additional information. A spectrogram though gives you an unbelievable amount of detail. For example, you can look at a voiceprint of a chicken and know it is a chicken right away. The human voice has a multitude of harmonics in its voice pattern and you can spot humans right away. So, if people are trying to hoax or fake something, it is easy to detect.

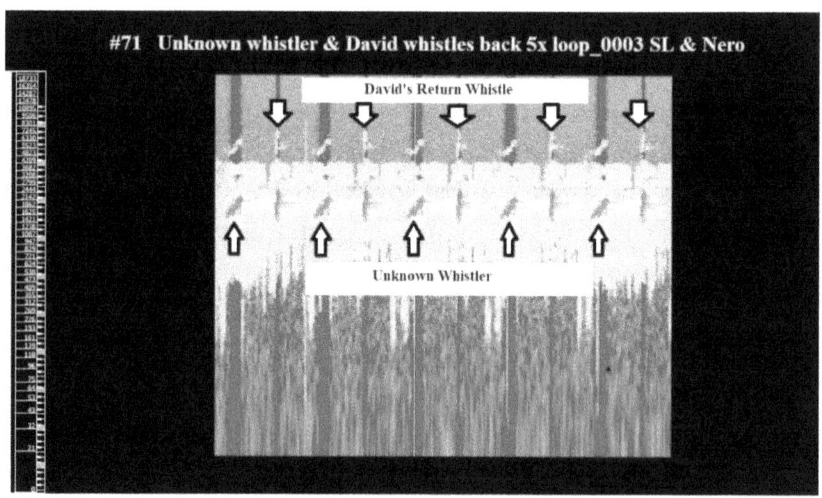

Sample Spectrogram, "Unknown Whistler" – Photo courtesy of David Ellis.

What keeps you going, what excites you about the subject? At this point, fumes, and coffee[laughing]. I guess it's been an obsession for so long. I love doing the analysis part of it, finding something new, and making correlations. For example, I may see two recordings with similar characteristics and these audio recordings are 3000 miles apart, they sound the same and they have the same suspicious nature to them.

Can you share one piece of evidence that you received that just really stands out? Daniel Boone was claimed to have shot a Yeahoa or Yahoo. So, one day I'm on the internet doing a search to see if I can dig up any more information on that. A YouTube video pops up that is called, *Wa-Hu, The Baboon Bark*. I say, "Wa-Hu? Oh, wait a minute. I got to look at this". I looked at the video, and it's a baboon, vocalizing what sounded like *wa-hu*, and I thought, "I've recorded that". So, I went back and I found one of the recordings here in the northwest that sounds just like that. Not only that, but it was also one of those vocals that had been recorded in Alabama, North Carolina, and Arkansas.

Ellis Sample Audio and Spectrogram

Do you feel since the 1960s that we have gotten any closer to proving the existence of Bigfoot? Yes, Dr. Jeff Meldrum's foot morphology is published and reviewed. Bill Munns's analysis of the Patterson Gimlin Film is published and reviewed. In addition, tens of thousands of eyewitness reports exist. DNA testing is becoming less expensive and may unlock some doors in the near future.

Do you think that we have enough evidence to prove the existence to the mainstream science? I think we do, but that's to the "knowers", the people that corroborate what they already know. Until there is a body, we will never have the evidence that will end the discussion. Currently, we are into the study phase of the subject. Since I already know Bigfoot exists, I think it's important to collect evidence in a fashion that science can deal with. That's what my goal is to turn over a mass amount of data so that when we have a collection of evidence that really points to their existence, which probably will include having a body, we can take science to the next level. There's a whole bunch of scientists that are right on the fence. The subject has been desensitized for approximately the last ten years. Mostly because of some of the content that has been on television and media. The stigma of Bigfoot is put in the same category as people chasing unicorns.

How to follow David Ellis:
www.olympicproject.com

DOUG HAJICEK

Doug Hajicek (pronounced high-check) is a recognized name in the Bigfoot community and his impact on Bigfoot research has spanned over 30 years. He also has had a prolific influence in various television and media genres. I witnessed Doug's great generosity firsthand during my first interaction with him. I asked Doug why Sasquatch: *Legends Meet Science* was not available on streaming applications. A few days later, I received a DVD version of the movie at my home, personally sent by him.

Doug Hajicek Introduction

Doug inspired me to make this book a reality, after sharing the book concept with him, Doug encouraged me to move forward with it, and in typical Doug style, we didn't waste any time! I've been fortunate to have the opportunity to interact with him much more during this project. He is truly genuine and supportive, and his creative input has been invaluable.

Doug has a brilliantly innovative mind and passionately acts on his ideas to bring them to life. This is evident in the groundbreaking scientific accomplishments he has had and the multiple successful projects he is involved with. Furthermore, based on the late-night/early morning emails and messages I receive; I am not sure if Doug ever sleeps. This leads me to one of Doug's most impressive qualities, he responds to every message, email, etc., always! I don't want to leave out his sense of humor though, he has a way to make you laugh without effort and typically while sharing a story about his personal activities or a past experiences.

Doug has certainly influenced science to consider theories on Bigfoot, he is driven to help solve the mystery. His documentary *Sasquatch: Legends Meet Science* continues to be regarded by many as the most impactful film to date on the subject.

Doug is a successful television documentary series producer and television show creator committed to bringing daring voices and innovative ideas to the screen. He is respected in the industry as a cutting-edge credentialed technologist, camera systems inventor, and wildlife researcher, documenting never-before-seen behaviors in animals such as black bears, beavers, fish, and invertebrates.

Doug is the president and founder of WHITEWOLF ENTERTAINMENT, INC., a television production company specializing in non-fiction programming. With over 206 features under his belt, Doug has been producing television shows for national networks since 1985. He is currently developing a full-length movie and/or series entitled: *Legend Meets Science II* or the *Legend Meets Science* series. The release date is planned for Fall 2022.

Doug's work has been nominated for three Emmy Awards, a Telly Award and a best product award. His long-running *MonsterQuest* series set many new ratings records on *The History Channel* and to this day is considered the gold standard for its genre.

See complete bio on link in book

Where did you grow up? In the sparsely populated wild suburbs of Minneapolis where streams, swamps and forests still existed. I never came into the house during the summer.

What are your hobbies outside? Electronics, engineering, wildlife photography, fishing, building: e-bikes, lightweight motorcycle building, graphic arts, exploring in general, traveling, filmmaking; not in any order.

Do you have musical talents? I played drums way-way back in the 70s.

Most memorable event and where was it? Seeing the hoaxed Minnesota Iceman at the state fair. There are too many others to list.

What are your favorite places you have visited? Austria, Iceland, Belize jungles and caves, Russia, Arctic Circle, Sea of Cortez, etc.... can't pick one as all unique and beautiful.

Can you share a favorite childhood memory? My first trip into the Superior National Forest and catching my first fish. It was on the Canadian border wilderness while visiting my great grandfather's cabin.

What was an embarrassing or funny memory? Flying and then falling off a high cliff while riding a dirt bike, I was trying to show off to a girl. I landed in a shallow river with no injuries, except to my pride.

What is one thing about your childhood that you look back and say, "what was I thinking"? Taking a cheap blow-up raft on a long float trip during an ice out down the Mississippi River with no life jacket... more than once. In warmer weather also.

What is your favorite seasoning for food? Pan-fried walleye or scallops in butter with adobe and blackening seasoning. Yum!

What is something that most people don't know about you? I have 6 great kids who are all brilliant.

What is a bucket list item that you would like to accomplish in the next 6 months? Start movie project - *Sasquatch: Legend Meets Science 2*, so I know the best documentary on the topic will get completed.

If you could take one person past or present on an expedition with you, who would it be and why? Someone who has never experienced true wilderness and would appreciate every second. There are also many experienced and creative researchers who are incredible people I would love to share a challenging expedition with.

When and why did you get involved in Bigfoot research? I love wildlife mysteries and seeing a detailed and classic tight rope 16-inch footprint trackway, first-hand in the arctic. There was one footprint in front of a stunted tree trunk and then the next one directly behind the trunk, I knew this was a massive creature and they were 100 percent real right then and there.

If you have had an experience(s), what was the most compelling one? While in a remote area, we experienced rock-throwing exchanges and had our cabin attacked 4 separate times. This was in a remote wilderness area after we heard sparse but loud wood knocking.

What was the most compelling witness account that you heard first-hand? I have heard so many hundreds of credible ones, I can't pick. A recent one is Jeff Harding's story on a sighting and face-to-face run-in with a Sasquatch in Canada.

What do you feel Bigfoot/Sasquatch are and why? I don't know. Most likely a flesh and blood- hairy and big human. Enough said.

What excites you about this subject? It's a mystery that hardly no one provides evidence, and no one studies it... for the most part. When sightings happen, most people clam up. When evidence is presented to the scientists, they clam up.

If Bigfoot is a large primate, what environmental conditions would it need to survive? Montane forests and access to food. It appears Bigfoot need 40 inches of annual rainfall or more, and access to water constantly.

How would you explain the elusiveness of Bigfoot? I can't, nobody can although I can make reasonable inferences: It is intelligent, rare, travels mainly at night, it is omnivorous and has a seasonal diet, and stays under the cover of trees, swamps, and deep forests. It migrates as food needs change. They are on the move on a regular basis. They get seen plenty, just not reported!

What are your current research goals and how do you go about them? I want to obtain clear, close-up sequential photos or video and baseline DNA accomplishments. Continue to think out of the box. Trying new things and new locations.

What do you feel is your strongest skill/talent that helps you in this field? Creativity is combining technologies – thinking out of the box and taking action, and plain old common sense. I am skilled and experienced in getting funding from TV networks.

If you could get better at one research-related skill, what would it be? Primitive trap-trigger skills that could be used to trip spring wind cameras. Tracking - one can never learn enough.

What do you feel is your biggest challenge that hinders your goals? The extreme intelligence of the subject in its home environment which we are pursuing.

What are your current projects?

- *Sasquatch: Legend Meets Science* 2 TV Film special pre-production
- ***Untold Radio AM Podcasts*** every Wed, where we discuss the Bigfoot Mystery - a lot

What is/are your biggest accomplishment(s) in this field? *MonsterQuest* has documented my accomplishments for the most part in Cryptozoology: The first giant squid filmed in its wild environment, the first mammals of different species living together in cohabitation and cooperation; muskrats, and beavers. I was responsible for filming the first wild black bear in hibernation. This led to documenting the first wild black bear birth and care of cubs filmed and resulting scientific paper co-authored. I created a laser animal measuring lens developed and used in the field by biologists above and below water. I was the first to film a 12-foot shark living in freshwater. I also developed and used the first *critter cams* developed. Furthermore, I was able to get the first-ever use of IR lighting tech for television. *Mysterious Encounters* series on the OLN network was the first TV series devoted to the Bigfoot mystery. I feel I have documented and shared Sasquatch research with the world before anyone else. This includes exposing and sharing with endless millions of people the science being applied to the Bigfoot Mystery. Proud to have instigated the first true complete scientific book written on the topic with *Legend Meets Science* - by Jeff Meldrum.

What is your elevator pitch for open-minded skeptics? The sightings will continue to happen, and the subject is fueled every day, and by hundreds of credible eyewitnesses a week and thousands per year. About 1 percent of sightings get reported due to skepticism. There exists good film, footprints, hair that doesn't match known species, dermal information, a body cast, and better DNA baselines are on the horizon. I suggest you do not ignore the topic as this mystery is not going to ever go away. It's a geographic problem, not a demographic one. Sightings seem to only happen near water, forests, and hills; and in most cases all three are present. That's my effective elevator pitch.

What are your must-read books for skeptics and new researchers?

- *Sasquatch-Legend Meets Science* by Jeff Meldrum
- *Apes Among Us* by John Green

What suggestions do you have for researchers who want to get involved in this field or further educate themselves? Talk to someone knowledgeable and ask questions. Listen to eyewitness reports and stories. Stay away from hoaxed social media Bigfoot footage.

Why do you think there is not more evidence or proof? I think there is plenty. Just not enough academic people studying it.

How do you feel mainstream scientists with publicly accept the existence of Bigfoot/Sasquatch? Future baseline DNA and new forensic technology that keeps getting better.

How to follow Doug Hajicek:
Facebook: Doug Hajicek
untoldradioam.com
whitewolfentertainment.com

CARLOS JIMENEZ

Carlos Jimenez may not be as publicly well known as others that we see in television and media regarding the Bigfoot phenomenon. His educational background is in the Sciences. He earned a B.A. in Anthropology at Pomona College under Dr. James McKenna. Carlos finished the Master's/PhD program that he started at Washington State University under Dr. Grover Krantz with an M.A. in Anthropology at the University of Wyoming under Dr. George Gill. His graduation project at Pomona-- where he also founded the Cryptozoology Club and was a member of the International Society of Cryptozoology, was on the existence of Bigfoot. At Wazzu, his unpublished thesis was about Gigantopithecines (very large fossil Asian apes that lived around 5 million years ago) and their hypothetical, ancestral connection to Bigfoot. During his time at Wazzu, Carlos was involved in giving Bigfoot presentations at the university and also at Sasquatch Days in Harrison Hot Springs, British Columbia, Canada.

Carlos has spanned the globe performing technical research, spending lengthy amounts of time in the jungles and wildernesses. This allowed for him to interact and learn from the indigenous tribes relative to the areas he was studying. He is open-minded to unexplained phenomena and is determined to find scientific explanations for them. As you may determine after reading about Carlos and exploring his projects, he diligently explores the historical data prior to hypothesizing his theories.

What is your favorite place you have visited and why? My favorite place that I have visited is Bwejuu, Unguja Island, Zanzibar, Tanzania. This remote fishing village lies on the east side of the island where I watched the *locals* harvesting octopus, rockfish, seaweed, and other sea life from the Indian Ocean. What I loved about this village were the seemingly endless white-sand beaches, the warm nights, the deliciously fresh seafood, and the overall beauty and tranquility. The stars filled the sky most nights, and during the full moon, you could almost see as well at night as during the day. The Jozani and Muyuni rainforests were not far inland, and there I saw myriad species of colorful and hirsute monkeys and all sorts of tropical plants and animals. Having grown up in rural Pennsylvania and lived in Los Angeles, this place was like no other I had ever visited in my life. Where else could you spend around $3 a day to have a tropical beach right outside your door and to be randomly visited by monkeys?

What is a bucket list item that you would like to accomplish in the next 6 months? A bucket list item that I would like to accomplish in the next 6 months is to host and create my first documentary film about Bigfoot. Specifically, I want to finish filming the interviews with Kirk Stewart and other native and non-native people who have had encounters with Sasquatch here in Del Norte County. I am looking forward to working with the director and producers to put together a thought-provoking and meaningful film sharing the various encounters that my protagonist has had over the past twenty years of his life off-grid, focusing on the main story arc of the time that he believes Bigfoot buried his gun. As a scientist, I love empirical evidence when it is available, and I am accumulating more and more photographic, film, and forensic trace evidence for Bigfoot the more I get out into the field. As an anthropologist, I equally appreciate the anecdotal evidence related to the Bigfoot phenomenon. I only want the beliefs and memories to be told by those who have actually witnessed these beings-- I do not want to tell someone else's story second-hand. I hope that these eyewitness testimonies and actual footage from the field will open the curious and/ or skeptical viewer's minds to the reality that we are sharing the planet with another, highly intelligent hominid species.

If you could take one person past or present on an expedition with you, who would it be and why? I would take Dr. Grover Krantz out with me. In all his lifetime and during his career as an Anthropologist, he never had his own encounter with Bigfoot. He always entertained my hair-brained ideas for different types of scientific studies we could conduct to advance Bigfoot investigations while I was his graduate student and teaching assistant at Wazzu. For the sacrifices he made to enlighten the rest of us about the scientific reality of Sasquatch, I would love to take him to one of the many habituation sites camping for a few nights. I would drive the Prius, and he could leave his helicopter at home (He had a helicopter in his front yard that he put together from a kit: I trusted it less than I trusted my Prius in the swamp.)

If they exist, what do you feel Bigfoot/Sasquatch are and why? I feel that Bigfoot are a hominid species. I infer this because of the reasonable observations witnesses have made of beings that exhibit bipedal locomotion, that have eyes positioned anatomically facing forward indicating stereoscopic vision, that display discernable homologous surface muscle structures to our own but are hairier (like our closest hominid relatives), that have opposable thumbs, and that many say appear human-like in their facial features. Additionally, reliable DNA results since the 1990s have consistently placed Bigfoot in the hominoid line but distinct from each and closer to us than the other great apes. Observed behavior such as ambush hunting and the construction of hunting blinds suggest something more akin to ourselves. Finally, research into their vocalizations has shown that beyond mimicry they also use morpheme streams indicative of language, and hominids are the only known animals that use language.

How would you explain the elusiveness of Bigfoot? I would explain the elusiveness of Bigfoot by discussing their hominid intelligence: the mental quality that consists of the abilities to learn from experience, adapt to new situations, understand and handle abstract concepts, and use knowledge to manipulate their environment. Witnesses have seen their wood structures and received items from them at "gifting" sites. Bigfoot are known to flank people, to distract them with sounds and noises to get witnesses to look at or in the direction of one of them, seemingly to allow others in proximity to relocate without being seen. They always seem to be one step ahead of investigators in the field and are rarely seen, heard, or otherwise recorded. Many hunters have noted that Bigfoot seem to know what a gun or rifle is and that their demeanor changes when they see the weapon. I posit that their eyesight is better adapted to night vision, perhaps being able to see infrared, that their hearing is superb, and that their sense of smell is better than ours– again, based on multiple eye-witness reports.

What type of research are you currently involved in? Currently, I am using camera and audio traps at a site where there have been several encounters over the past twenty years. I am collecting trace evidence from this location, including footprints, hair samples, unnatural structures, and scat. I am examining the hair and categorizing it, making permanent slides of samples of the known and unknown species. I am planning on getting software and analyzing the visual patterns of the audio from recordings we are making. I am hiking and camping at the location whenever possible. I have been invited to join several other expeditions in 2022, and I am excited about those opportunities. Working on a team is something I have not had the pleasure of doing yet, and the more eyewitnesses the more credible the report.

What researchers past or present have influenced your work? Researchers who have influenced my work include Dr. Grover Krantz, John Green, Dr. John Bindernagel, Dr. Jeff Meldrum, Peter Byrne, Kirk Stewart, Doug Hajicek, Les Stroud, Jason Wells, Kevin Merriman, David Ellis, Derek Randles, Cliff Barackman, Amy Bue, Aleksander Petakov, Will Jevning, Russell Acord, Mireya Mayor, Tim Baker, and many others whom I am not remembering or who would rather not be named.

Do you feel since the 1960s that we have gotten any closer to proving the existence of Bigfoot and why? I do not feel that we have come any closer to proving the existence of Bigfoot since the 1960s. Technology has gotten better and our population has increased, so the number of people in the wilderness with a camera in their hand has increased. But I am not certain if there has been a change in the statistical significance of the quantity of anecdotal evidence or trace evidence that has been amassed in the past six decades one way or the other. What I can say with some confidence is that we are becoming more open-minded about disclosure as a society. Even the military and government are starting to release classified or denied information regarding UFOs (now UAPs) as more and more boomers are coming clean on their deathbeds about the secretive and classified work that they did in this lifetime. Many feel a moral obligation to let the public know the truth. And guess what? Society did not go into a panic and start running amok. Perhaps these divulgences will eventually snowball and lead to the disclosure of other secretive, classified programs and the information that they have discerned and withheld from the public– like perhaps the existence of Bigfoot or extraterrestrial biological entities. I am not holding my breath on this one, as I still think that admitting that there are unknown aircraft/ spacecraft/ seacraft– with the preponderance of drone technology– is a far cry from the who-or-what is-piloting-them disclosure. If one of us came across a dead bigfoot body in the wild or along a roadside or train tracks, I still do not believe that we would be allowed to share this "proof" of Bigfoot's existence with the general public. I might be wrong... I hope I am wrong.

What are the Chronicles of Carlos? For my brand, the Chronicles of Carlos, I investigate, create, and share mainstream educational content, extraordinary Bigfoot encounters and other mysteries, mythologies, and legends of the unknown, and examine and discuss the scientific methods and principles corroborating our stories. My platforms are for a very large audience of those curious about Bigfoot as well as the Bigfoot knowers, believers, and skeptics.

What are your current projects and how can people follow your work? I am currently making my documentary film "Bigfoot Buried My Gun." Next, I will be working with Doug Hajicek on a yet-to-be-disclosed collaboration. Folks can find information about my film and follow or contact me on any of the following:

How to follow Carlos Jimenez:
CJ@ChroniclesofCarlos.com
ChroniclesofCarlos.com
YouTube: Chronicles of Carlos

PHOTOS AND INFORMATION

Color Images of PNW Bigfoot Maps

Josh Moss is an expert Esri GIS software analyst. He utilized that skill set to create PNW Bigfoot Maps. The data is compiled from Bigfoot reports from 14 diverse North American reporting organizations.

Bigfoot Report Statistics
Total – 8,522 reports in 49 states and 9 Canadian Provinces
Top 3 States – Washington (853), Oregon (713) and Pennsylvania (620)
Runner up – California (602)

Busiest Decades/Decade Range in database – 1811-2016 (200 years of reports!)
1970's – 1,345 reports
1980's – 1,137 reports
1990's – 1,681 reports
2000's – 2,719 reports
2010's – 819 reports

Seasons
Spring – 1,281 reports
Summer – 2,723 reports (Busiest)
Fall – 2,347 reports
Winter – 1,050 reports
Unknown – 1,121 reports

(Citation for raw data) Mangani's Bigfoot Maps
http://penn.freeservers.com/bigfootmaps/

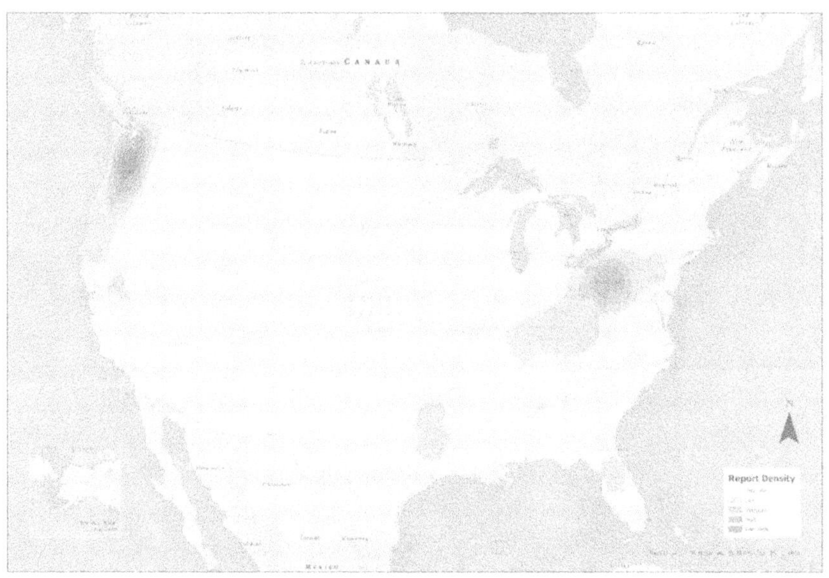

Bigfoot Sightings Heat Map. Photo courtesy of PNW Maps.

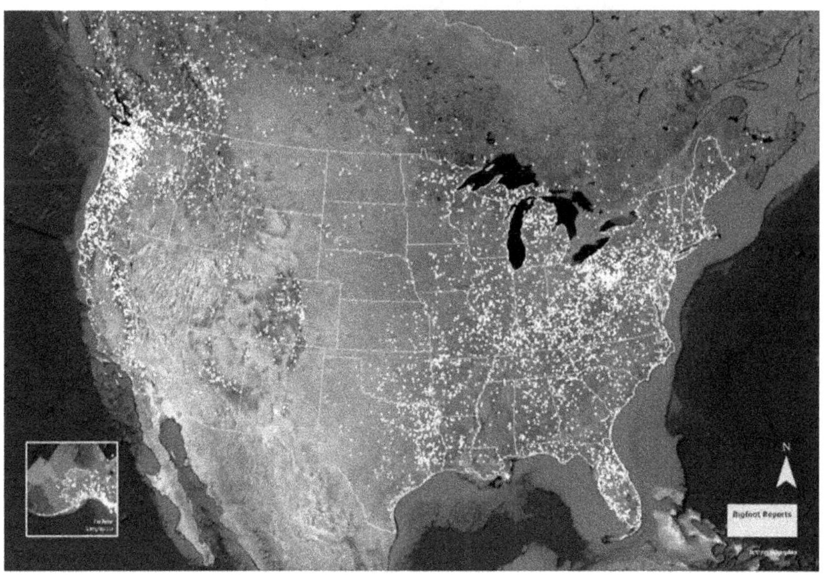

Bigfoot Sightings Map. Photo courtesy of PNW Bigfoot Maps.

Artist Snuffy Destefano with his chainsaw carvings. Photo courtesy of Snuffy Destefano.

Author Lyle Blackburn's Book Cover. Photo courtesy of Lyle Blackburn.

Dana Halloran with artist Bo Bruns creation. Photo courtesy of the author.

Bob Strain, Bob Gimlin, and Kathy Strain. Photo courtesy of Kathy Strain.

Operation Sea Monkey in 2016, the team included Thomas Steenburg, Ron Morehead, Gunnar Monson, Thomas Sewid, Todd Neiss, Victoria Williams, and Darren O'Brien. Photo courtesy of Todd Neiss.

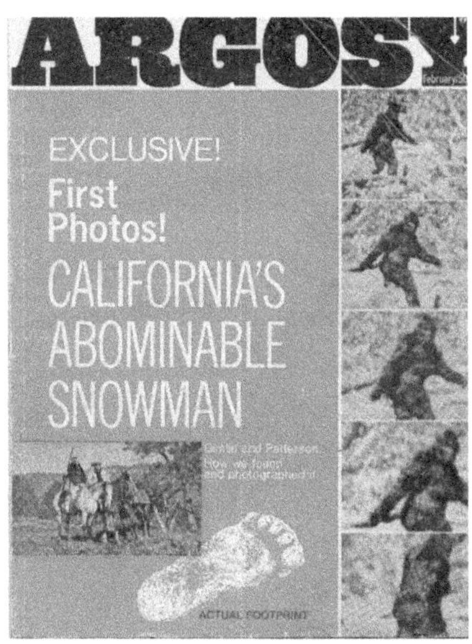

The images of the iconic Patterson-Gimlin film were first featured on the cover of the February 1968 issue of Argosy magazine. Photo courtesy of Todd Prescott.

Peter Byrne examining a squashed mouse which was found at the bottom of a Sasquatch track. Photo courtesy of Rick Noll.

René Dahinden and Rick Noll at Harrison Lake. Photo courtesy of Rick Noll.

DR. RUSS JONES

I met Russ a few years ago at the Ohio Bigfoot Conference. Instantly, he welcomed me with a smile. Russ was very approachable. In addition to his warm personality, by his appearance, it was evident that Russ was also focused on his physical health. Getting to know Russ has given me the opportunity to understand that he is one of the nicest and most kindhearted researchers I have met. In addition, his knowledge related to Bigfoot is impressive and his approach is data-driven based on tangible environmental and wildlife activity that he has collected and reported for the past three decades. Russ is truly a kind and passionate person. The Bigfoot research community is better because of his humble and genuine investigative approach.

Dr. Russ Jones Introduction

Russ grew up in a family that raised cattle and farmed. This type of environment encouraged him to grow up outdoors. He spent much of his time in the woods hiking, hunting, trapping, and fishing. He graduated from Huntington University in Indiana where he was on a baseball scholarship. After earning his bachelor's degree in science, Russ received his doctorate degree at Palmer College of Chiropractic. He has been practicing in Charleston, West Virginia for 30 years. Russ is also a certified Master Naturalist and a Master Gardener. He has appeared on television including *Finding Bigfoot* and has authored the book, *Tracking the Stone Man*, documenting his Bigfoot research in West Virginia. Russ is presently finishing up his second book, *The Appalachian Bigfoot.* Furthermore, Russ is a member of the Bigfoot Research Organization (BFRO) covering reports in West Virginia and Southern Ohio.

Where did you grow up? Southern Ohio/West Virginia

Hobbies outside of BF? I am a Master Gardener. At one point I had 50 rose bushes, and I was a member of the Charleston Rose Society. I was planning on competing in shows. Unfortunately, the new property I moved to in West Virginia does not allow roses due to the high deer population in the area. I also enjoy farming, being on the tractor, and in the past my family raised cattle. I am interested in anti-aging practices, and physical exercise and I work out regularly. I participate in anti-aging seminars, which also crosses over into my profession as we help our patients live healthier. Some call it "bio-hacking", which is essentially manipulating exercise, vitamins, and nutrition to obtain specific goals from your body. As a doctor, it is important to help guide my patients in taking care of themselves, how to raise immune function, and how to live healthier.

Bigfoot is a significant part of my life; I am not sure if there are many of us that do something Bigfoot-related every single day, however, I am one. If I am not in the woods, I am interviewing or talking to someone about the subject.

I typically try and get a couple hundred hours in the woods every 6 weeks. Some scientists have documented that for approximately every 200 hours in the woods, you typically have some type of encounter (sighting, tracks, vocals). I have found that for me it is also accurate. I am fortunate as I have gotten older and my professional practice has grown, I am able to get out in the woods often. I have now found as a result of more time in specific locations the 200 hours have come down.

Favorite musicians or bands? Wow, I tend not to listen to much music. I am more of a talk radio/podcast person. Generally, I am diverse in my music though. I like college sports. I am a big college football fan and a diehard Ohio State Football fan. I have been a season ticket holder for 20 years; I go to almost all the home games. I listen to *all* Bigfoot podcasts, literally.

Most memorable event and where was it? Ohio State Football games and Bowl games. The past Ohio Bigfoot Conferences, connecting with my friends and researchers.

Favorite Place you have visited? It is interesting. I have traveled to many places. My desire when I do retire from my profession is to take an RV and Jeep and travel to all the historical Bigfoot places of significance in the United States. I would like to explore the landscapes and meet some of the people that I may have communicated with in the past and never met in person. The reality is, my favorite place to be is out in the woods Bigfooting. I want to help solve the mystery.

Favorite childhood memory? My favorite memories are all the times I have spent in the woods with my family. In particular, with my grandfather, who I was very close to. I am sure I drove him insane (laughing) because of the hours and hours we spent in the woods hunting - I would shine my flashlight on every track we would come across and ask him what they were.

I would like to just have a day to tell him everything I know now about Bigfoot and to have his opinion. I bet that once he knew all the evidence that was out here, he would reflect on his many experiences (he was in the woods his entire life) and possibly have new insights on the things he experienced. I remember on multiple occasions when we were nighttime hunting with our lights off, we would hear *knocking sounds.* He would say there were just car doors closing and the sound got distorted as it carried through the hollows. He wasn't trying to fool me; that is just what he thought it was. He may not have realized that there could have been something else out there making those sounds. Expanding on that, to have an experience, you must know what is out there and what it looks or sounds like. Prior to *Finding Bigfoot,* most people in the public did not even know that Bigfoot make sounds. Many people may have heard wood knocks and other sounds and just passed it off as something else just as I and my family did in the past, simply not knowing.

What was an embarrassing or funny memory? One time I was with a group of researchers trying to find a way across a river. We had looked around for a couple of hours when we saw children walking on the other side where we wanted to go. We yelled across and asked how they got there. There was a bridge path we didn't see because of snow. The story has morphed to now being children in wheelchairs, but it's a fun memory.

What is one thing about your childhood that you look back and say, "what was I thinking?"? I don't really have any regrets. When I was young, I wanted to grow up and be a preacher. When I got into my undergraduate program, I changed and went into the sciences. I then eventually continued to earn my doctorate degree. The past has helped to define the path I have taken and where I am now. My sisters would say that they were horrified of the short running shorts I would wear [laughing]. In the 70s & 80s, I ran marathons and other events. I would wear the typical running shorts of that time and they still joke about it.

What is something that most people do not know about you? I'm a Master Gardener and enjoy writing poetry on occasion.

What is a bucket list item that you would like to accomplish in the next 6 months? To get my second book wrapped up.

Author update: Russ' book, *The Appalachian Bigfoot* has now been published.

If you could take one person past or present on an expedition with you, who would it be? Bob Titmus. He was an outdoorsman in Bigfoot research starting in the 60s. He didn't care about the spotlight; he just went about his business. He was an expert tracker and cast many footprints.

When and why did you get involved in Bigfoot research? I had a couple experiences when I was younger which got me interested. After my second experience, I watched *In Search Of*, with Leonard Nemoy, and got suspicious of my previous experiences. I believed them to be Bigfoot-related. A couple years later, my dad and I were in the woods and found a footprint and my interest expanded.

While in college, I stayed interested and read many books about Bigfoot. I was consumed with it and eventually went to the Oho Bigfoot Conference. At that time there were only a few hundred people who attended. Dr. Jeff Meldrum was speaking at the event for the first time. I spoke to him after he presented, and a few months later I decided to participate in a local BFRO expedition. I told myself, "If these people are weird, I will leave. If not, I will stay". I met

Matt Moneymaker on that expedition. We hit it off, and I started doing the BFRO reports in Ohio and West Virginia.

I am just as consumed with the subject now as I was back then. It is interesting how events that happen to you when younger can shape your life. My encounters and interests impacted where I bought my farm, and where I chose to practice, all built around my ability to get into good Bigfoot territory near where I live.

If you have had an experience(s), what are the most compelling one(s)? It was New Year's Day. I was 15 and out with a friend who was in the army. We were hunting rabbits. We had gotten a few inches of snow the day prior. It was cold, maybe 15 – 20 degrees and sunny. We were traveling along a hillside where a cave was located. This cave was unknown to most, almost hidden unless you knew the area like we did. As we approached, we found a fresh set of foot tracks. We assumed that something had weathered the previous storm in the cave. Whatever it was must of heard us getting close with the dogs and left. At that point, I knew nothing about Bigfoot. We looked at the tracks and I thought it was a barefoot human that may have sought refuge in the cave. We searched the cave and there was no sign of human occupation.

Approximately 6 months later, I was fishing in a beaver pond which was in another remote area. We knew about it from hunting the area for so long. This pond was approximately 1 mile back into the woods not a likely place for human traffic. Furthermore, this was in a remote small county. Most of the land is state or national forests. I was with my uncle fishing and 40 yards from the other bank. It was very brushy so I couldn't see very well. Both of us were wearing pistols because of the many snakes that lived in that area. I suddenly heard something traveling down the bank across from me in the brush. My uncle was 50 feet from me and was also watching the area that the noise was coming from. The bushes started shaking and we heard approximately 60 seconds of heavy animal breathing sounds that to me sounded monkey-like. Then it stopped and we continued fishing.

We hunted and fished our entire lives and weren't worried. We were raised that there wasn't anything in the woods that we should be afraid of; everything was afraid of us.

Approximately how many eyewitnesses have you spoken with? Approximately 1000 different ones, although I do follow up on reports and speak with witnesses multiple times. Daily, I check the BFRO Ohio, Kentucky, and West Virginia reports.

What was the most compelling witness account that you heard first-hand? My 1st BFRO case was the most compelling. It was a state trooper and his wife. They were riding their 4-wheeler in the summer scouting for possible areas to hunt ginseng the following fall.

He was traveling on a right-of-way - I see multiple reports in these types of areas - they are typically long straight passageways and not near homes. In addition, a right-of-way provides an edge for other animals to travel like deer, game birds, etc... In this account, the state trooper had taken a turn from the right-of-way, and he stated that he glanced up and saw a "fire burnt stump". He then glanced down and then when he looked back up again, he paused and said, "have you ever seen one?". I said, "no", and he said, "they are the size of a sheet of plywood, they are so large, it is hard to imagine that something like that could be in the woods." It was leaning back behind a small tree 20 yards from the witness. He was wearing a gun although it was so enormous, he did not even think of pulling it out. As he is trying to get the 4-wheeler in reverse to retreat, his wife, who was on the back of the 4-wheeler asked, "what are you doing?", he said "look, look, look", and then she said, "oh my God, no!". They did get turned around and tore out of there. He told me later that it was interesting and almost funny that the Bigfoot was so large and was hiding behind such a small tree. His wife later had to get counseling for post-traumatic stress, and they decided to move from the country to the city. Even though they moved, every nighttime noise she would hear outside made her fearful that Bigfoot was in their neighborhood.

I later brought the witness back to the site with another investigator. While there, he was clearly shaken, he was chain-smoking and sobbing. Furthermore, another state trooper accompanied him. They both had their guns drawn the entire time. I knew he truly saw something.

What do you feel Bigfoot/Sasquatch are and why? Well, it would be easiest scientifically if it was Paranthropus or Gigantopithecus as they are part of the record, and the path makes sense. I think in likelihood it's a relict hominid. It's fair to say something from Eastern Ape lineage.

What excites you about this subject? Honestly, I find it odd and mysterious that this hasn't been solved. I have personally interviewed multiple hunters that have had clear daytime sightings while hunting. Each one of them could have shot one, although they didn't attempt to. They weren't anticipating encountering a Bigfoot in the woods and it only happens for a few seconds and then it's gone. They seem to have human-like characteristics and it catches you off guard.

There are good reports documented in my area on a regular basis. Of those reports, maybe once a year you will get an exceptional report, a daytime sighting from a credible witness (ranger, doctor, police officer, etc.). Those compelling reports excite me.

In addition, I listen to hundreds of compelling legitimate witness accounts on podcasts which allows me to hear repetitive behavior traits with Bigfoot. There is enough happening to you and others in the woods that you know this is realistic and it keeps you going.

If Bigfoot is a large primate, what environmental conditions would it need to survive? Anywhere that deer thrive, and where black bears exist would be a good start. Usually, places with the most rain have many sightings. They need enough room to have security and enough room to be able to meet their caloric intake. On the East Coast, they would need maybe 600 to 1000 square miles.

How would you explain the elusiveness of Bigfoot? I think people primarily discount how few of them that there are. They aren't everywhere and it can be hard to get near them. They are curious about humans but not actively seeking them out to interact with [them]. They have lived in the area for generations and know where people go and where they don't.

What are your current research goals and how do you go about them? I want to be able to accumulate enough data that I can reasonably predict where they are at a given time. I want to be able to get a clear game camera photo, though for scientific reasons that may be a challenge. Ultimately, I want some of my findings or hypotheses to help verify the species.

I keep a Bigfoot calendar which helps me document activity based on location and time of year. I have found that since I have been tracking data on the calendar, it is not taking me 200 hours to experience something in the woods. The data allows me to be more strategic and encounter activity much quicker(sic). The calendar includes specific dates of witness sightings/encounters, my encounters, and historical activity in the geographic areas I am interested in.

When I started researching, I traveled within a 6-hour range of where I live. I had cameras set up in those areas as well. As time went on, I recognized I was not being efficient. So now I am only focused on investigating 3 unique areas which are closer to me, and it has allowed me to be much more efficient. I have eight to ten cameras in each of those areas. I feel the activity of all the wildlife in those areas is important as well. For example, if I see the same deer show up every day in the same area, and then there is a period when they don't show up, there could be something suspicious happening at that place and at that time. I add this to my calendar. In addition, if I have reported Bigfoot activity in a particular area and at a particular time noted on my calendar, I will hike in that area at the same time of year. Sometimes the witness reports may not be accurate, so I must use caution on the data I add.

What do you feel is your strongest skill/talent that helps you in this field? One of the skillsets that helps me is that I have had well over 200,000 patient visits. You get a real feel when you have seen that many people regarding what you believe and what you don't believe and what is the truth and what isn't. In addition, I have hunted, trapped, and ran dogs my entire life and am also a Master Naturalist and have spent so much time in the woods. Since I have been Bigfooting, there are certain things that just don't make sense to me and there are others that make it a little easier for me to understand based on my knowledge of the woods. For example, if you are experienced in hunting mushrooms, morels to be specific, in the winter you start recognizing areas that look viable for the next season. Then you go back and look for them in that area when the mushroom season is in. I feel you can take the same logic with looking for Bigfoot. I do feel though, there are more potentially good habitats for Bigfoot than there are actual Bigfoot.

It seems a common occurrence that Bigfoot are after berries. During berry season, I am looking for right-of-ways that are remote and don't have easy access for 4-wheelers. Normally I have also found that they stay on north-facing hillsides during warmer weather. I utilize *Google Earth* to investigate further in these types of areas. In addition, Bigfoot don't seem to stay in the main hollows. It is the side hollows where activity is. How this helps, for example, I can look at a 10,000-acre forest and quickly analyze if the topography matches what I am looking for. If it does, I then will walk the hillsides multiple times to see if I can find anything.

In the past, at any given instant, if I would get a compelling report, I dropped everything and investigated. It has taken me a long time to learn not to "chase ambulances". Recently, there was a road sighting of a juvenile Bigfoot crossing the road. This was witnessed by individuals in 2 vehicles, and they did not know each other. It was compelling and I did follow up; I called both individuals. I did realize though that it was not efficient for me to drive the 2 hours to investigate because the area of the report is like many other environments in West Virginia and based on experience, I would not find any evidence.

If you could get better at one research-related skill, what would it be and why? An under-utilized skill in Bigfooting is tracking. There are many good classes and programs available. I have taken some. It comes down to making time to get in the woods. Many researchers can only get out in the woods a few times a year and that can make it challenging.

What is your elevator pitch for open-minded skeptics? When I wrote my first book, I did a couple pages on what potentially Bigfoot could be. Most commonly researchers feel they are Gigantopithecus or Paranthropus. I could have written many pages on each theory. Instead, I felt that if I laid out the most credible evidence and some compelling eyewitness sightings that at the end of the book someone might just say, "You know it seems reasonable that this is something worth checking out". We all know that witnesses are notoriously unreliable when it comes to height, hair color, etc. They all are perfectly capable though of realizing and reporting whether something was a human or not a human. We also need to weigh the best compelling reports that we receive from people who are trained to observe (doctors, rangers, law enforcement, etc.) and consider these more credible.

I don't think that it is reasonable to me that hoaxers would be going across remote locations in North America and planting fake evidence where no one may ever find it. *Bergman's Rule* is that generally the warmer the climate and the closer to the equator you are, a species is typically smaller in body surface and mass, whereas in colder climates they are larger. This is true for found tracks from potential Bigfoot, the farther north of the equator, the larger the tracks are. How would hoaxers know to do that on a regular basis? When you weigh all the potential evidence, there is at least a chance that Bigfoot are a living species today. For those of us who spend the time researching this on a regular basis, there is little question in our minds that witnesses are seeing something. Retired Yellowstone Ranger, *Bob "Action" Jackson*, who supported wildlife conservation and spent his career deep in the wood (traveled nearly 70,000 miles Yellowstone's backcountry), had a Bigfoot sighting. He certainly knew what animals were native in that region.

I also feel many people have opinions on the subject although they aren't into Bigfoot or well versed in the subject. Many have not spent any time in the woods and still have an opinion on Bigfoot. If Bigfoot is not one of your interests or hobbies, it is not likely you would know of all the compelling evidence that has been collected.

What suggestions do you have for researchers who want to get involved in this field? First, spend some time with people who are interested in Bigfoot. If that means taking a private or BFRO expedition, do that a few times to get experience in the woods with individuals who know what they are looking for. I also suggest getting training (tracking, naturalists, etc.). There are plenty of good programs available and they allow you to learn about the woods and the environment. I sometimes see social media posts around tree breaks, some seem falsely related to Bigfoot. If you don't understand the environment, you cannot be certain what caused the breaks. They could be caused by disease for example. Approximately 80% of the tree breaks you see on social media are in pine areas. Pinetrees are notorious for weak roots, they sway in storms and have much surface area to catch wind. I do carry a mini microscope with me so I can instantly examine potential evidence if needed.

Find a place that has potential or a history of Bigfoot activity and is close enough where you can get there often, at least every week or two. It doesn't do any good to pick a spot that is 4 or more hours away where you can't get out in the woods that often. You don't get out enough to learn the area. You can install game cameras, understand what type of people visit, what animals are in the area, and when they are moving through. The areas I work in, I have learned about the wildlife (bobcats, deer, etc.) and their habits as well as how much human activity there is. That is all part of learning an area.

When I first enter a new area, I walk all the trails and then I walk all the creeks (it may take me many months). After that, I walk all the ridges in the area and install game cameras in traditional areas. I may also install a camera in a location where I can monitor how many people are traveling through at different times of the year. I typically leave the cameras in for at least a year. I have a few park rangers that I communicate with regularly. They ask me questions about the wildlife activity, human traffic, and other occurrences. They don't have the opportunity to get in the woods often, so I provide them important information about activities in particular locations at various times of the year.

Why do you think there is not more evidence or proof? I feel it is the rarity of the animal. I think there is evidence out there, although we are going to need a body or body part. We are collecting hair samples, blood, eDNA, and nests; in time these will mean something. The primary issue is the lack of animals, again the rarity of the animal we are dealing with. For example, in West Virginia, I estimate we would have between 150–225 Bigfoot. That is not a big number. Think about each county which may be 25 square miles or so, maybe there is a small family of 4 in each county. These animals have learned to live there their entire life avoiding us for the most part and have learned where humans travel and where we don't. For example, if you are a hunter and lose your dog, it is hard enough to find them and they are drawn to people. We are talking about something that seems to be inquisitive about people although generally stays away from them. We have 22,000 bears in West Virginia; I only get approximately 5 sightings a year and I am in the woods all the time.

Another problem is that we just don't have researchers out all the time in the woods to get evidence. Around the world, there are scientists getting full-time funding to search for a species. Unfortunately, we don't have that in Bigfoot research. Some scientists have hundreds of cameras set up in specific regions and go years without getting footage of a *known* species they are looking for. We have part-time "citizen scientists" that are devoting their time, getting cameras out, although they just don't have the resources like funded researchers.

It is also possible that Bigfoot can hear the electronic noise of the cameras. Another example: we know that it is scientifically rare to get an image of an alpha male coyote for the same reasons.

Some ask, "where are the bones?". If there are between 2,000–20,000 Bigfoot in North America and if they live an average life span like a gorilla or orangutan, we can estimate they live approximately 40 years. With a 5% annual mortality rate, we would only have a maximum of 300-400 dead Bigfoot to find a year. Breaking those numbers down by state/region, you can see why we don't find bones. For example, in West Virginia, we may only have 3 Bigfoot die per year, which will make it almost impossible to find bones. I do get forestry service professionals that I connect with. We discuss unexplainable noises they hear and they mention that they don't find many deer bones relative to the high population of them (estimated at over 500,000).

In addition, I don't go out at night anymore. I have learned that we don't accumulate compelling evidence at night. The thermals are fantastic. Unfortunately, the scientific community does not see it that way and they aren't interested. I used to spend so many nights alone in the woods with my thermal; you hear things and catch some convincing visuals, although not enough compelling activity though to push the science forward. I also think people, in general, overestimate their ability in the woods and underestimate the ability of an animal in the woods ([chuckling] "that is my quote of the day and I have never said that before"). Some think that every noise they hear in the woods are Bigfoot. The reality is Bigfoot is rare and they are not everywhere. If they were not elusive our forefathers would have seen the signs of the animal and hunted them down in the past and we would have scientific evidence.

How do you feel mainstream scientists will publicly accept the existence of Bigfoot? We are going to need a body or a body part. Sometimes I think the internet does not help; scientists who may be interested in the subject may get discouraged by seeing the abundance of inaccurate information that is being shared on social media. Unfortunately, there are many nonscientific images, videos, audio, and reports, that are shared on social media which may not be related to Bigfoot at all.

How to follow Dr. Russ Jones:
www.thebigfootdoc.com

STEVE KULLS

Steve works as a Licensed Private Investigator, in New York., and has had a lifelong fascination with the Bigfoot phenomena since an early age. He has authored three books related to the subject of Bigfoot, "50 Large," "What Would Sasquatch Do?" and "The Sasquatch Playbook". He has also appeared on Fox News, and the History Channel, the National Geographic Channel, Travel Channel and Destination America have featured him on several programs. He has also appeared in print in over 200 newspapers worldwide.

Steve has interviewed over 500 witnesses related to Bigfoot encounters and numerous individuals in his profession due to his 30-plus years as a private investigator. He is a trained forensic interviewer which uniquely qualifies Steve to vet out the truth from hoaxes. Many know Steve as the "Squatchdetective" which began as a profile name for his email when he started researching in the early 2000s and has stuck ever since. You can follow his work at Squatdetective.com or his popular YouTube channel, Squatch-D TV, where along with his colleagues, Steve debunks hoaxes and misidentifications related to potential Bigfoot evidence. His show is no-nonsense when it comes to getting to the truth while keeping the mood fun and entertaining.

Although Steve focuses some of his activity on investigating potential hoaxes, he and his research team are actively in the field looking for signs of the elusive Bigfoot species. He has played a vital role in helping enthusiasts and other researchers weeding out hoaxes to uncover the truth.

What type of research are you currently involved in? I am involved in the behaviors of the creatures. Why do they do what they do and how it falls into the realm of primate behavior and areas of habitation.

What researchers past or present have influenced your work? First and foremost, Bill Brann. Followed by Dr. Warren Cook, Paul Bartholomew, Tom Steenburg, Dr. John Napier, Dr. Grover Krantz, Dr. Jeff Meldrum and finally John Green.

If they exist, what do you feel Bigfoot are and why? Some sort of undiscovered primate. Now whether it's a hominin or pongid, or something in between you got me. But a primate.

How would you explain the elusiveness of Bigfoot? Their defense I believe is avoidance. Primates are forward and abstract thinkers that can plot their next move. I think that is part of the reason they are so hard to find. Them finding you is more like it.

What was the most compelling witness account that you heard first-hand? John Whitesel, who sadly passed away at an early age in 2012. He was an avid outdoorsman, with many wild animal encounters. He was with his wife hiking, and they decided to travel off-trail and ended up next to a stream. Suddenly, something close by started to loudly scream at them. With all of John's outdoor experience, he had never heard anything like it before. The sound was so guttural, he could feel it vibrating in his chest. He and his wife decided to get out of the woods quickly. While hiking out, they could see something paralleling them, moving from tree to tree. It was approximately 100 feet from them, they could see the dark upright figure moving along with them while it continued screaming. They were frightened and it followed them for approximately ¼ mile until they got back onto the trail and then it ceased. This was an amazing and incredible account because it was not the typical brief encounter, it sustained for 15 to 20 minutes.

Why are credible witness accounts important? First, by the enormous numbers of eyewitness accounts that have been reported. People may throw out that, "witnesses aren't the most reliable people in the world". Considering I have been a private investigator my entire life, I deal with a lot of witnesses. For example, if a bank robber runs out of a bank and I have five people that see him, I will get five variations of what that person looks like. There is one commonality, they all see a human, it is not a gorilla or a bear walking out of the bank. We can still say that a bank robbery occurred and generally witnesses are good at identifying if the individual was a male or female. The bottom line is when it comes to eyewitness Bigfoot reports, you can take the fine minutia and argue against certain characteristics although they all have major commonalities. Maybe the most important is that what they saw was an unknown creature with primate features and certainly not another animal like a bear, human, moose, etc...

What was the most compelling experience or evidence you have encountered firsthand? I was with my research team camping in the same area that John Whitesel had his sighting. It was the evening; I had left the campsite temporarily to go back to my vehicle for new batteries for my headlamp. After I grabbed the batteries, I decided to shine my flashlight down the dirt road to scout the area. I then swung the flashlight uphill and saw a large creature standing next to a utility pole. I could make out the silhouette and the long hair although not much else. My speculation is that it was observing us while we were by the campfire. When I put the light on it, it turned to me, then looked behind itself and turned back to me. We both froze, I could see these two eyes glowing back at me and blinking. After approximately 30 seconds, I shook the light a little and the creature turned and went back into the woods. I immediately radioed the team. They could hear it moving behind the base of the camp although we did not get another visual.

What do you feel motivates individuals to create Hoaxes? My professional career requires me to study the behavioral traits of people. When people create hoaxes, they generally do it for one of the following reasons. First, you have the "joker" who is just looking to get a "haha", they are typically easy to point out. The second group is what I call the" profiteer", someone who is trying to make money off the hoax, Ivan Marx for example. The 3rd and largest group are people who have a phycological need to feel important or need to elevate themselves to feel special.

What is your elevator pitch for open-minded skeptics? I am not out to convince anyone that Bigfoot exists, people will believe what they want to. If you have an open mind, keep it because I have seen and experienced things that I cannot explain and I am a grounded person. I always invite people to come out in the field with us and if they do, they may experience something that may change their minds.

What suggestions do you have for researchers who want to get involved in this field or further educate themselves? Set your expectations low and realistic, and have fun. I also suggest inviting a significant other or a family member to join, let them feel part of the experience as well or at the least have a nice fun camping trip, don't let it put a wedge in your family. Additionally, don't believe everything you see and hear, always cross-reference your facts and always stick with people you trust (it may take some time). Stay grounded, always tell the truth, and if you are wrong, you must correct it.

How to follow Steve Kulls:
Squatchdetective.com

DR. JEFF MELDRUM

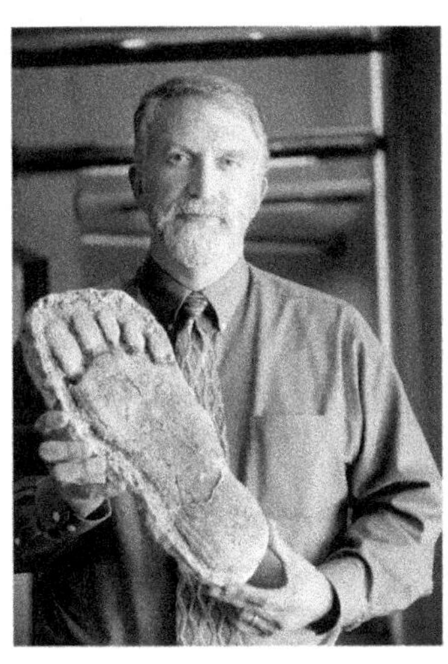

Dr. Jeff Meldrum is currently the prominent academic involved in Bigfoot Research. He is a professor of Anatomy and Anthropology at Idaho State University, focusing on primate locomotor adaptations and modern human bipedalism. Being an expert on primate bipedal locomotion, much of Jeff's research on this subject is focused on footprints; specifically, his lab houses over 300 compelling casts believed to have been made by the creature known as Bigfoot.

I first met Jeff at the Ohio Bigfoot Conference, where we happened to be checking into our rooms at the same time. It was a day prior to when most arrived, so it was relatively quiet. I could see Dana at a distance in the corner of my eye awestruck seeing Dr. Jeff, who is her favorite Bigfoot-related personality. I then jokingly said to Jeff, "You kind of look familiar, do I know you?" We laughed and then throughout the weekend, I had the opportunity to have multiple conversations with him. Most of, if not all, of our discussions, were not related to the subject of Bigfoot. Getting to know him personally, rather than as a media star, was truly enjoyable and a highlight of the conference. Further, going into the weekend, I was skeptical about the existence of Bigfoot. However, Jeff impacted me by explaining the complex and involved efforts the individuals were putting into research. After hearing his lecture in person for the first time, I thought, 'Wow, there are legitimate scientists studying this phenomenon and Bigfoot may actually exist.' My dialogue with Jeff fully opened my eyes to the possibility.

The baton for Bigfoot research was passed to Jeff by the late Dr. Grover Krantz, who previously was the academic face of Bigfoot. Dr. Krantz laid the foundation for scientific research on the subject. When Jeff was featured in Doug Hajiceck's documentary *Sasquatch: Legend Meets Science* and subsequently authored the complementary book for it, there was no turning back. He became the "go-to" scientist on the subject for the media, enthusiasts, and researchers. Although there are no experts on Bigfoot, Jeff is an expert related to his academic specialty, which directly associates the possibility of the existence of the creatures based on locomotion and footprints. Today, he continues to carry the academic torch with the goal of providing evidence for the existence of Bigfoot to the world. Jeff may be the most renowned individual in Bigfoot research, and without his presence, the science would not be where it is today.

Where did you grow up? I grew up all over the Pacific Northwest. My dad worked for Albertsons, so we were transferred quite regularly, every two to five years. I was born in Salt Lake, I lived there, Provo, Portland, Eugene, Spokane, Boise, Walnut Creek, California, back to Boise and then went off to school back in Provo at Brigham Young University. I then did my graduate work at SUNY Stony Brook on Long Island, I did a two-year postdoctoral fellowship at Duke University in Durham, North Carolina, and then my first academic posting was in Chicago at Northwestern. We lived just outside of Chicago - in Fox Valley, Aurora, Illinois.

What are your other hobbies? My work really is my hobby. I'm quite a jack of all trades in some ways. I don't specialize so narrowly as so many in this highly technical age do, so my attention is diverted in many different directions in my academic field. It doesn't leave a whole lot of time between that and a large family. I think my number one hobby was going to soccer and lacrosse meets every weekend, sometimes multiple, on a given Saturday with all the kids. Likewise, I was also a scoutmaster for eight years, given we had six boys. That was another weekend that was taken up with camping, which also gave me the opportunity to be out there with my boys. We had a lot of fun doing that. I've always loved hiking, backpacking, a little bit of hunting and fishing, outdoor photography and again, jack of all trades, not really a master of any, but enjoying dabbling.

How do you feed all six of them? Well, we got creative and the grocery bills were always a bit of a shock, you bite the bullet and set your priorities. It was funny though, one day my number two son Sean came up to me and he said, "You know, dad, the other day I was over at the in-laws and we had orange juice for breakfast, and I couldn't figure out what was wrong with it, it was so sweet". To pour eight glasses of orange juice before emptying the pitcher was a challenge, plus it is too sweet for my taste. I would always add another can of water, which made it better and stretched it a little further. He said, "you were watering down the orange juice, weren't you?"

There were never leftovers. I heard this figure recently that American families throw away 40 percent of the groceries that they buy. Not in the Meldrum house! [laughing] We were quite resourceful and used the most perishable items first in the front end of the week. There were leftovers, but there was always a leftover day towards the end of the week and we cleaned out the fridge before going back to restock. We've just always been very efficient and the kids were not picky eaters, some were a little more finicky than others. It was always fun to experiment with those who are more willing to try new dishes and things, we were just efficient and economical.

What's your favorite type of music? I'm kind of eclectic. When it comes to favorites, I have a hard time pinning things down. I usually just get on YouTube and bring up something to let it play for an hour. I've been listening to slow jazz and enjoying that recently; it is a new spin for me. I like acoustic guitar, Jesse Cooke for example. When I need a little more rousing in that after lunch hour, I sometimes put on things like Two Steps from Hell to hear epic battle songs to get the blood circulating and brain cells going. My wife and I are a little different, she's a diehard classic rock fan and can name off the bands and all their members. I can rarely associate band names with the title of a song. Being eclectic, I can tolerate and enjoy it with her, on a small scale. Similar to the TV, I'm the type who's always flipping channels, I'm turning the dial or pushing the buttons on the radio to hear some variety. I can't listen to very many performers' entire albums, it's too monotonous. I like mixes, the variety.

What's the favorite place you visited and why? I've had the good fortune of having the opportunity to visit numerous places. The one that sticks out is the trip to China with MonsterQuest. China was just so different. I lived for two years in Germany and I have some German ancestry. My maternal grandmother was a first-generation American from a German lineage, so many German traditions, cooking habits and so forth were familiar to me. I went to China and the food and culture were so unique to me. You can recognize the basic needs, the basic human expressions of living life, but they've done it in such a different way. I was quite tall by comparison to most there. The first time that someone was transfixed by the hair on my forearms was really quite funny. One person could not resist the impulse to just literally reach out and pet my arm, he was fascinated by this hairy barbarian. I thought that was quite amusing. They were excellent hosts and showed me around and exposed me to the different cultures and foods. We had an opportunity to see the beautiful landscapes, some of the exotic geology and forests and so forth, and different plants and animals. That was quite a fascinating experience, so it's always high on my memory list.

What is something that most people don't know about you? I might say that I have a sense of humor. Sometimes I come across in interviews as rather stoic and contemplative and I've had several people comment that they were surprised that I was just kind of a down-to-earth, easy-going, fun-loving guy and could even crack a joke occasionally. This story came to mind thinking about this: Derek Randles had invited me out to the Olympic Project to speak for one of his retreats, and one of the things he often does is give a demonstration of basic survival gear, so you're safe when you go out. His rule, which he repeated over and over again throughout the demonstration, was that you have those essential items with you, especially that top five; a blade, a light source, a fire source and so forth and you keep 2 of each, so you always have a backup. So later in the day, Derek was complaining because he had run out of Tabasco sauce, and from the side of the room, I just very casually and dryly said, "You mean you didn't have 2?" The whole room just burst into laughter, and what they thought was even more funny than the joke itself was who it came from. Derek says, "Yeah, watch out for him. He is just like a snake in the grass there waiting for a chance to strike".

Who's your favorite Star Trek character? Oh, it's got to be Mr. Spock. I'm less Captain Kirk, you know bucking the prime directive and more the logical course of action to take, the scientific side. As I've gotten older. I guess a certain degree of cynicism comes with it, I hope that's not a sign of early-onset senescence. I tend to be more rational, more logical, less emotional, less sensational.

If you could take one person past or present on an expedition, who would it be and why? The first person that immediately comes to mind is the person I spent the most time on expeditions with, and that's John Mionczyski. John is just a phenomenal individual. It's hard to describe him, and you really can't appreciate him unless you have spent time with him. He's just this walking almanac of knowledge regarding natural history and botany. He's a wildlife biologist and ethnobotanist. He also knows the native cultures and traditions, concerning this topic that we have shared together, and he has provided such an interesting insight, both scientific and the cultural/social side of it because of his close affiliation and interaction with tribal peoples, especially over in the Wind River Reservation of Wyoming. You'll never be hungry if you're out there with John because he puts together a salad bar with ease based on what's available, it's really quite fun. Hopefully, some of his information and aspiration for understanding nature have rubbed off on me.

When did you get involved in Bigfoot research? There were a couple of blips in the timeline, and the first, obviously, was my first contact with the subject matter, which was the Patterson-Gimlin film, when it was shown in Spokane, Washington. I think it was either 1968 or 1969, we've had a little trouble pinning that down. I've got the newspaper advert, but unfortunately not the dateline from the page it came from. Then, that interest waxed and waned over the years with the competition with other topics, other avocations and interests. Then as an academic, in a position to better appreciate the significance of the evidence itself, it really was the interacting with the footprints that I was shown at Five Points outside of Walla Walla, Washington. There were a few events that kind of primed me for that experience -- an invitation to evaluate a piece of video footage that was shot in Northern California, which dusted away some of the cobwebs and reanimated some of the interest and enthusiasm for the subject. Then sort of a spontaneous, impromptu visit to Dr. Krantz's Laboratory and spending an afternoon with him and spreading out the casts and talking shop, literally. On the return trip the next day, my brother and I paid a surprise visit to Wes Summerlin and Paul Freeman in Walla Walla. That was it, that really set the hook, it was 1996. I've been very actively investigating, pursuing, and evaluating since then.

Your brother was involved as well? He was at that point.; I had gone to Boise. I don't remember what the occasion was, some three-day holiday or something. We decided to spend it at my parents' in Boise for a little change of scenery. I remember discussing with my dad and my brother the experience that I had evaluating this piece of video footage from Northern California. Years prior, the three of us had gone to the Patterson-Gimlin film airing in the Spokane Coliseum. So, it was kind of fun, we were reminiscing, and of course, I had a long-standing interest in the subject immediately after that. The interest got packed away as I got older when things like girls and dating became more important.

We were sharing recollections, and I was telling them about this trip to California and this curious piece of video footage that I could not find the zipper, to point to a hoax and what it might mean. Our wives were heading out with the credit cards, and I've always wanted to go up to Grover Krantz's lab in Pullman. It's not that far from Boise. I turned to my brother and said, "You know, let's take a little road trip, you and I will go do an overnighter, we'll drive up there and visit him, and then see what he actually has". So, we did that, and it was a fascinating experience. Grover was very collegial and hospitable and spent several hours with us, and I had my video camera, one of those giant ones that sat on your shoulder. I was videotaping and asking questions, and we were laying out these casts and talking about the anatomy and Grover's interpretation, and my impressions. It was quite amazing.

So, backing up to my previous trip to Northern California to evaluate the site of the videotaped encounter. I was there at the invitation of Richard Greenwell, who was the secretary of the International Cryptozoology Society. On the way home, he handed me a book from his briefcase, which is a little paperback book called *Bigfoot of the Blues* by Vance Orchard. Vance was a journalist in Walla Walla. I think at the time he wrote the columns for the Waitsburg Times, and a couple of other regional papers that would run his outdoors column. He frequently devoted those columns to recent events on the Bigfoot subject, things that he heard from Paul Freeman, Bill Laughery, or Wes Sumerlin, these were characters that are in the Walla Walla area. Anyway, Richard asked me to write the book review, so I was reading it on the plane and when I got home, and right up to the weekend I decided to see Grover. I'd had a previous phone conversation with Paul, who extended an open invitation to visit him. I'm thinking as we're leaving Pullman, the roads were a little bit icy and we had driven up through McCall on the highway instead of the freeway and had witnessed several cars that had slid off the road.

I said, "You know, it's even worse today, let's take the freeway and on our way, we'll make a pit stop in Walla Walla and stop in to see West Sumerlin and Paul Freeman", unannounced, just take our chances. So, we did and had a great time with Wes and then we drove over to Paul, he was just pulling into his driveway. We introduced ourselves and he immediately recognized my name and invited us in and was very hospitable. He generously started pulling out boxes from the closet and was laying these casts out on the floor. Man, it was something because these were originals, they weren't like the copies I was looking at with Grover, which were high-quality copies, but when you look at an original, there's a little something different about it, it has something, it captures the essence. I don't know what it is about the original, it just has a different presence.

I was pointing out specifics and asking questions, pressing him on details, straining his recollection of circumstances because this represented over 10 years of accumulation of evidence. That's when he said, "Well you obviously know a lot about footprints. Would you like to see some fresh tracks?". I said, "What do you mean?". He said, "I found some just this morning". Every spring, as the snows would melt off the foothill roads, he would start driving these roads and up to the ridge lines. Outside of Walla Walla, there are a lot of dirt roads, literally just graded dirt roads with no gravel bed, there's a high concentration of Pleistocene loess, this very fine powdery sediment in the soil that makes for great tracking. It's just like walking in wet clay.

To make a long story short, that was how the hook was set, and it was set firmly by the quality and quantity of this evidence.

Which current or past researchers have influenced your work? Obviously seeing Roger Patterson, in person and hearing him introduce the documentary showcasing that 60-second clip was very impactful. Shortly thereafter, I got a hold of John Green's book, *On the Track of the Sasquatch*, that had a huge influence as well. The first book I had ever asked Santa for as a Christmas present was Ivan Sanderson's *Abominable Snowman Legend Come to Life*. I devoured that repeatedly, that was how I gained an understanding and appreciation of geography and vegetation zones across the globe. I came to love maps through his discussions of all these exotic areas and figuring out the distribution, and how mountain ranges influence climate and such. Then I was delighted when John Napier published his book and the influence that had on me, another scientist. Eventually, Napier would have a significant influence on my professional career because he was sort of the father of studies of primate locomotion and was a physician who specializes in hands and feet.

Obviously, Grover as well, he was never really a mentor per se because we just didn't interact even early on as anthropology and anatomy students. I remember my first contact with Grover. I must have caught him at a less than opportune moment, we were at the Physical Anthropology meetings and one of his graduate students, Don Tyler, who later became a faculty member at the University of Idaho, had done a paper on a species of New World primate. At that time, I was getting more immersed in New World platyrrhine paleoprimatology and ecology. As we were chatting, Grover came into the auditorium. Don said, "You ought to go say hi to him, introduce yourself". I walked over and I introduced myself and I said, "I just wanted to let you know what an impression you have made on me because of your work with Sasquatch". He says, "Sasquatch, that's all everyone brings up is Sasquatch, what about all the other work I've done?". I just felt like I had a bucket of cold water poured on me.

Later I got to know Grover, he was very nice and when I approached him on more equal footing, perhaps. We started dialoguing long before he retired and sharing information and insights, about dermatoglyphics, about my developing model of the Sasquatch foot, which differed from his significantly and fundamentally. I was gratified in his second edition, he actually acknowledged the insights that I had added and acknowledged the likelihood of a midtarsal break, and midfoot flexibility that differed from his interpretation of the foot and the footprints. Then, of course, when he discovered his cancer and that his life expectancy was extremely curtailed, he was very happy that I was in a position and desirous of taking the baton from him. There was someone to carry the torch and there was a home for his cast collection and his notes.

I ended up competing, though, with the Smithsonian, which expressed an interest, so he made a very selective subsample of some of the casts and molds and some of his correspondence, which went into their special collections. The rest he earmarked to come to my lab for me to continue working on. I immediately got a call from the collections manager at the Smithsonian, introducing himself and getting acquainted and wanting to make sure that when I reached retirement if I would consider the Smithsonian as a permanent home for the remainder of Grover's collection. I don't know if that's the best place because it's kind of like the scene in *Raiders of the Lost Ark*, with artifacts disappearing into the warehouse. For someone who has the credentials to gain entry to the hallowed back halls of the collections area, they can be examined. Although, I think they would be much better served, perhaps somewhere like Cliff's Museum, for example, if there isn't another academic, a young academic. I'm going to hold out hope that within the next five years or so that someone emerges as a rising star, if you will, in the academic eye.

Do you think when Grover first met you, he would have thought that you were the guy he was going to pass the baton to? You can see it when you read his book and get to that final chapter, where he calls it the "scientific establishment", he had become very anti-establishment and rightfully so. I've experienced the same thing, where you have an abstract submitted, and it pivots on the opinion of two people, basically, whether it gets accepted or not. Those times that they have been accepted, the reception, and the discussions that ensued have been extremely productive. Then on the times that they've been rejected and I've pushed back, there is no appeal process. I still push back and I rattle people's trees over it. You get silly responses like, "Oh, this topic is of no general interest to the anthropological community". It's just an inane rationalization, so I can understand his sour grapes in the end, he was a bit jaded and quite cynical. As I said, cynicism tends to percolate as you deal with people with narrow minds and in positions of power arraigned ambitions or egos.

Primatologist John Napier published his book, *Bigfoot, Yeti and Sasquatch* in 1973. Do you feel his position may have changed now based on what we have learned about evolution since then? I think maybe. It's interesting in his book, and I think justifiably, he sort of came down negatively on Yeti as a probability. He did clearly acknowledge that there was something to Sasquatch. He said something is leaving footprints. It was because of his background in anatomy, his specialty and hands, feet, and locomotion that he was impressed by the footprint record. Even though he misread it, he didn't have enough variation, a large enough sample, and came up with this imposed dichotomy between hourglass and human shape.

Unfortunately, now in retrospect, his book is not very useful, it's good as a historical snapshot of the thinking, and it was at least useful that an academic went out on a limb, so to speak. He felt he could afford to at that stage of his career. It's interesting when I talk about the Patterson-Gimlin Film and his assessment in the book, he admits he really can't put his finger on why he thinks it's a hoax, except that when he looks at that thing, he sees from the waist up essentially an ape and from the waist down human proportions. He said it was almost impossible to conceive of such a contrivance, such a hybrid in nature, so one or the other had to be contrived, hence it's a hoax.

Well, his book came out in 1972 and then in 1974 we had the announcement of the discovery and naming of *Australopithecus afarensis*, Lucy.

When I do my presentation, I have this image where I take the top half of a bonobo and the bottom half of an African pygmy and stick them together, and then I point out, when you read the popular statements, the public statements in the popular press by anthropologists about Lucy, how did they describe her? From the waist up she looks like a chimp from the waist down she looks like a little hairy human. They will go on to say how interesting it is that evolution has combined traits in unexpected ways. Well, my question is, what if his book had taken 10 years to write instead of 5 and it wasn't published until after that public awareness of a very different perception of early hominid evolution? Would it have affected his opinion? That was the only thing that was the linchpin of his statement on the Patterson-Gimlin Film. Now we know that combination is not counterintuitive, it's absolutely in line with the norm for early hominin functional morphology. I think it would be fascinating if you could roll back the clock and reexamine some of those things.

Can you recall an embarrassing or funny moment while you were researching? It was on that China trip. On that trip, I had the good fortune of traveling with Adam Davies, one of the cast members. Adam is probably 6'2', he's fairly tall. Between the two of us, we were kind of imposing figures amongst the much smaller stature of the Chinese. One of the witnesses that we dealt with was, Mr. Yuan, who it turns out, had a pair of fascinating footprint casts. He prided himself that he was 5' 10", which made him a good half a head taller than anyone else in the village. The first thing he would do when he'd meet you is reach out and shake your hands and pull you right over, he'd sidle up, side by side and measure. He was rather chagrined that not only Adam and I but quite a few of the camera crew were also quite tall. Given that context, we were very limited in what we could bring with us for this extended trip. Of course, camping equipment really wasn't in the offering, so that was provided by one of the local fixers.

[Laughing] When we had hiked into the Shennongjia Nature Reserve with our small entourage, we were getting ready to set up camp and they handed us a tent. Adam and I were going to be tent mates, we pulled out this tent and it's a little pup tent. It was even small by Chinese standards; we pitched this thing up thinking our feet were going to stick out through the door. Thankfully, one of Adam's duties was to man the blind overlooking a spring that they had baited with some fruit. So, he spent much of the night into the wee hours of the morning watching from this blind. But when he joined me in the tent, there just really wasn't a lot of room. I tell you the floor of the tent was so narrow that we couldn't fit both, and as a result, it pushed down the side walls and pulled the arch of the tent down. I remember at one instance during the night, I woke up and I opened my eyes and there's his face. I'm not kidding, our noses were almost touching. So, I rolled over to the other side, but of course, as soon as I would roll over, the tent would start to bend and it would lift on the other side. At one point, Adam elbows me and says with a heavy accent, "Jeff, move over, you're squashing me". It was funny.

What's the most compelling evidence you've encountered firsthand?

Jeff Meldrum

What one piece of evidence do you feel that mainstream science can't discredit? They have failed really on all the enduring evidence like the Patterson-Gimlin film, and the composite footprint evidence. Some within the scientific community may think that it has been discredited, but it really hasn't. I'm more confident now about the Patterson-Gimlin film than when I was as a 10 or 11-year-old back in the late 60s. That confidence is based on data and analysis, not just first impressions. It's interesting that the academics who have been most impressed, who have taken the most serious look at the evidence, it has been the footprint evidence, whether it was Napier, Grover, or even George Schaller, it was the footprints and the Skookum Cast that really piqued George's interest and curiosity. There is just so much.

I had a department chair who sort of berated me and said, "Well, after all, Jeff, these are just stories". He was of the mindset being that he had been born and raised in Idaho, he hunted and fished all across the state, and if there was a Bigfoot out there, he would have bumped into it, he would have found it, seen it, or encountered it. I said, "Well, just stories that leave footprints, that void scat apparently, that shed hair that can't be attributed to any other wildlife, that vocalize, that are seen by credible eyewitnesses, and have a distinctive presence in the archaeological record of the Native Americans". One of my colleagues chided him and said, "Have you ever set foot in Jeff's lab? Have you ever looked at his footprint collection? As a scientist, how can you make a decision without examining the primary evidence first?".

It's this attitude of, "it can't exist, therefore, it doesn't exist", and it doesn't matter what evidence you think you may have. That was said in response to my pushback when a solicited paper was rejected by the scientific advisory board, of the journal of the California Academy of Sciences. Greenwell and I had written an article in response to their request for sort of a state of the science of Sasquatch. We put together, I thought, a very nice summary and acknowledged where there were gaps and acknowledged what still needed to be done but advocated very strongly that the evidence more than justified a serious, objective consideration of this subject and not an offhanded dismissal of it. It got the kibosh from their scientific advisory board.

If they exist, what do you feel the Bigfoot are and why? Let me make this very clear, there are alternative hypotheses that I think are very worthy of consideration and I always get a little bit annoyed when people label me as, "Oh, he thinks it's just an ape". It's just like just saying, "It's just a man in a fur suit". It's the most misplaced use of the word "*just*". These people don't understand what apes are, and the marvelous creatures that they are. It boils down to just basically which side of the split between hominins and hominids, in general, you want to advocate, and I don't think there's quite enough evidence yet to make that distinction. So, it's kind of silly to divide an audience with that. On the one hand, it has long been advocated that *Gigantopithecus* is a reasonable candidate, and I agree it's the right size, at the right place, at the right time. We only know a little bit about it, limited to jaws and teeth, but you can infer much about its morphology and its diet, and therefore its habitat and so forth from that. We don't know for certain if it was bipedal. If someone came out of a cave in Vietnam with a giant femur of a bipedal primate, then boom, we've got it nailed.

On the other hand, when I talk about the Paterson-Gimlin film, I often show a juxtaposition that I've composed for a slide with a *Paranthropus boisei* skull. When you line that to the absolute scale, point for point some things stand out, the very unusual cranial proportions: the exaggerated face, the lack of a frontal slope - a forehead going nearly straight back to a high point posteriorly on the skull, the very deep jaws, the flaring cheekbones, etc., point for point it lines up. You could infer some of those similar proportions and adaptations in a *Gigantopithecus* if we had more complete cranial material. Grover's reconstruction is just that, it's a speculative or inferential reconstruction based on extrapolation, imposing some assumptions about its brain size, that it is a large-bodied ape and therefore had a brain on par with other apes and or early hominins. But the similarities are quite remarkable, and when you realize that *Paranthropus* persisted in East Africa until just under a million to 800,000 years ago, *Homo erectus* was already evolving along the path towards us, with this other creature besides us, hence the name *Paranthropus*, parallel to man.

Although, it's the wrong size 5 ½ feet tall, in the wrong place, East Africa, and at the wrong time. That has always lessened the probability of that for me. It would have been much more likely to view the similarities that I see as convergences to a similar morphology in *Gigantopithecus*, Sasquatch, and *Paranthropus*. Then came the discovery of the Hobbit, *Homo floresiensis*. It's clearly a late australopithecine early *Homo* like *Homo habilis*, it's not an insular dwarf form of *Homo erectus*, as was originally suggested, just given the timeline and the juxtaposition, it clearly has greater affinities to *Homo habilis*. The other dot to connect is clear over in East Africa, two million years ago. So what happened in between? Obviously, there's a lot of chapters in that book that we are missing. And so that also leaves the possibility that something like *Paranthropus* could have spread out of Africa right alongside a *Homo habilis*, which ended up in Flores and the *Paranthropus* may have spread along southern Asia and up into northern climates and achieved gigantism, along with Gigantopithecus and other Pleistocene megafauna.

I think both of those hypotheses are reasonable. I draw the line, though, at *Homo*, there's nothing about Bigfoot that suggests *Homo*, one of our criteria has always been the manufacturing of stone tools, "Tools make it the man" as Oakley said. Now we do have evidence of simple flake tools, just knocking off a sharp little flake of stone that can cut, like a razor blade can cut through hide or cut off tendons of the meat from the bone as early as 3 to 3.5 million years ago, australopithecines were doing that, apparently. The point being, you look at Sasquatch, you look at Patty walking across that screen, she looks like a very early, robust australopithecine, grown to a bigger size. She doesn't look like even a Homo habilis, she's not carrying stone tools. We don't find little cobbles and flakes scattered around the Pacific Northwest except made by *Homo sapiens*. She doesn't have control of fire. She doesn't have home bases, she doesn't have clothing, she's not carrying the spear, she's not carrying a club, etc.

Why do you feel we humans have developed differently than any other primate? It's intelligence, tools, material culture, and language. The acquisition of language changes the way your brain functions. The way in which their brain is wired and functions is different. Imagine what your thought process would be like if you weren't able to talk to yourself, to articulate thoughts and make plans and so forth, it has a profound impact.

How would you explain the elusiveness of Bigfoot? I think people are sometimes too quick to attribute to Sasquatch extraordinary intelligence based solely on its ability to avoid human detection or acknowledgement. I mean, obviously, we detect them. We have sightings, encounters, and footprints all the time. So that argument isn't really a substantial one. Any time you talk to a hunter, ask them about the buck that got away, that prized buck that has survived season after season, and how? It has learned what it needs to do to survive and so it doesn't take human equivalency to outsmart us, to avoid us, to remain elusive. The combination of rarity and being smarter than your average bear, as Yogi would say. Also inhabiting a terrain that we are not adapted for; steep, rugged, broken, thickly vegetated environment. That is a challenge for humans to navigate. Yes, we can climb, although we don't have the endurance that other animals have that live in those environments. I'm always struck by an eyewitness who will comment on the fact of how amazingly rapidly this creature receded or climbed up a slope. There was one who said, "it would have taken me forty-five minutes to get up to that ridge top. And this thing did it in ten".

What excites you about the subject? It's a mystery. For me, science is equivalent as much as anything to exploration. The scientific method is sometimes applied by people in my perception, who are somewhat mundane. When you already know the answer and you're just trying to confirm that you know what the outcome is, that's not nearly as appealing as something that's on the frontier of science, as sometimes it's called, the edge of knowledge and understanding. Not to diminish the other, but different strokes, basically, and I prefer to push that boundary a little bit further. That's what's exciting to me, it is the unknown. I mean, there's many layers, and levels of involvement from the appeal of scientific inquiry. There certainly is a mystique about something out there that we haven't discovered yet specifically, an unknown creature, because I've always been fascinated by nature and animals, especially rare and elusive ones. There is also a sensational aspect to it that lends a certain thrill to being part of that search, being in the outdoors and engaging with it firsthand.

What impact do you think your books had on the field of research? I hope it has introduced to an open-minded audience the fact that there is a scientific context. This isn't just stories, it's legendary, folklore, it's myth versus science. Basically, the bottom line, the premise of that book is posing the question; Is there a biological species that can be examined scientifically behind the legend of Sasquatch? If it opens that door and holds it open, then I think it's successful. Establishing that scientific context is important and I've even matured in my perception of that since the writing of the book, now I appreciate even more the importance of the temporal context of scientific perception and framework. I think we can understand the perception by the scientific community through time, by understanding what the prevailing ideas and paradigms were influencing the discipline at a given time. In short, in the 1960s, the only concept that was primarily discussed and had a lot of influence on the study of human evolution and the interpretation of this growing fossil record of hominin species was what was called the single species hypothesis.

The idea largely borrowed from ecology and ultimately from microbiology, that no two species can occupy the same niche. It was also advocated by people who I think thought themselves a little progressive, more inclusive. They would have been very comfortable with some of the abused verbiage of today. "We're all humans, one happy village, one happy family". This arbitrary distinction of species is just imposed, "We're all just one happy family and have been for the past seven million years since we got up and started walking on two legs". Basically, the idea being that no two species can occupy the same niche that was the human niche. So, there could be only one species in it and evolution was just this gradual change, even if you acknowledge different species, one was supplanted simply or transitioned into the other smoothly in this lineage. This hypothesis did not allow for any non-human bipedal hominin species to coexist alongside us. When Ivan Sanderson's book was published in the early 60s, it wasn't even given a chance, skeptics thought he was just making up stuff and they can't exist, therefore they don't exist, it doesn't matter what you've marshaled in your book as evidence. That persisted until at least into the 70s when the tree began to explode with diversity and it was inescapable, the acknowledgement that there were multiple species coexisting at any one time.

To understand that is critical to understand why some won't accept the Patterson-Gimlin film. There's no place to accommodate the appearance of a bipedal hominin on that screen living alongside us. Even though within a few years, Richard Leakey demonstrates in East Africa four species coexisting, early *Homo erectus*, *Homo habilis*, *Homo rudolfensis*, *Paranthropus* all in the same landscape. Then the corollary of that realization is the acknowledgement, as exemplified by the discovery of the Hobbit, of many of these branches on this bush that has persisted alongside us until much more recently than ever would have been imagined. You could go back in time in East Asia, get out of your time machine and you could potentially bump into any of a half a dozen different species of bipedal hominin looking across the landscape. So why would we think that just in the last twenty thousand years, things have changed completely and we are the only hominin on the face of the planet when there's so much evidence that would argue to the contrary?

It's all in my book, we're ready for volume two. And in fact, Doug Hajicek is underway now with *Sasquatch: Legend meets Science, II*. I've already been informed that there's the expectation I will write the companion volume to go with that as well, so it's a great opportunity to do a second edition with new lines of evidence, new developments, and I'll be able to incorporate more of this information.

How are you spending your time or what are you currently doing, you know? What are you involved with your research right now? Well, especially in the wake of COVID, a lot of the fieldwork and travel has been curtailed until recently, and so I've devoted more attention to writing and editorial work. This past summer was quite productive and one of the projects that I've been working on in collaboration with a cultural anthropologist at the University of New Mexico at Gallup is *Bigfoot in the American Southwest*. We've received some exceptional testimonials shared by members of the Navajo Nation, the White Mountain Apache and other tribal peoples. That lends a very interesting ethnographic insight into the phenomenon and a biogeographic perspective on the ecology and biology of that region.

I'm the editor of the *Relict Hominoid Inquiry*, we're in our eleventh year. This past summer, I collaborated with a very talented artist to produce a set of learning and activity books for young readers on relict hominoids. Each book is devoted to a particular hominoid, they have puzzles, mazes, word searches, coloring and drawing activities. I cover the historical aspects, some of the movers and shakers, also the geography, maps, ecology, ethnography, the natural history of wildlife, tracking, and so forth. It's just been fun and gratifying.

What are the most compelling traits of the footprints evidence that just are undeniable? There were two ways to answer it. One, there are traits about the nature of the footprints, namely their remarkable consistency, their biomechanical appropriateness, those aspects that are extremely compelling. The repeat appearance of recognizable individuals in a geographical area. That was one of the first questions I asked as I determined to pursue this academically was, "If these things exist, then they must be very rare. If they are that rare, then if footprints are found repeatedly in a given geographical area, the chances of there being footprints of one of a few recognizable individuals repeating their appearance should be quite high". One of the things that always struck me is that it seemed that every footprint that was pictured in a newspaper clipping or in a book was unique, it seemed was distinctive, and it only was with greater familiarity and increasing the sample size that it quickly became clear that, yes, there are some superficial idiosyncrasies, but the underlying architecture and the relative proportions, the distinctions of toe proportions and toe configuration and lack of an arch and heel breadth and midtarsal flexibility, all those things come through in the credible examples, remarkably consistently. So that was quite a realization to be able to demonstrate sometimes to the counter intuitiveness of it. In one case, I would show two examples side by side and say, "Look, these are the same individual", and people look at it and state, "What? No, that can't be".

I explained, "Ignore the overflow, allow for the splay of the toes and they're the same, look at the shape of the big toe. Look at the way the little toe, the angle and its shape and notice the length and breadth are all the same if you allow for the difference in the depth of imprint or the overflow of plaster". Cliff Barackman does a fantastic presentation, where he has examined tracks from the Walla Walla area and at one point, he puts up a slide and there are literally about 18 casts and they're all clearly from the same individual when you start looking at them closely. When you look at the extremes of variation, you might not be inclined, although when you can show the intermediates of the appearance based on variation in those traits that I mentioned, you can see the continuum from one to the other. There's no way to draw a line between them anywhere and state, “OK, these will go in this bin, these go in that bin". It makes a very persuasive case.

I think there's both. Both of those aspects are very intriguing and compelling. The nature of the data itself and then the specifics of the architecture and biomechanics exhibited in the foot, which is very aptly adapted to the manner of locomotion. Witnesses describe the compliant gait and the correlation of the evidence of the footprints and the dynamic signatures and features like the metatarsal pressure ridge that can be observed. This is one of the things that makes the Patterson-Gimlin film such an amazing package. I tell people if all I had were the footprints that have been documented from that film site, just the footprints, I would be convinced of Roger and Bob's story. Then to have the film where I can see the trackmaker and watch the foot doing the very things that I've inferred from the appearance of the footprints and see how that correlates with all the rest of the anatomy, its size, its bulk, its compliant gait, its arms swing and so forth. You just tie up the whole package and put a bow on it. It's just amazing.

Jeff Meldrum

Are you finding more scientists open to the possibility of the existence today versus when you started? Yes, I am somewhat, and it's been very gratifying recently. There have been several examples of young PhDs just out of their programs, just hitting the job market, who have expressed their interest in Sasquatch. I kind of tell them, "shhh" for job security because we're on a cusp of a transition. Thomas Kuhn described paradigm shifts, he coined the term paradigm shifts and very insightfully acknowledged that sometimes when there's this turnover of ideas of concepts in a discipline, it may take an entire generation passing before the new paradigm can take root and find free expression. We're still at a point where even though the shift has taken place, the generation immediately before me is still in the old mindset. See when I was a graduate student, in the early 80s there was still talk about the single species hypothesis. There were still adherents to that notion that you could trace humanity back millions of years, not hundreds of thousands of years. So that shadow cast its influence far forward into the future. Those of my generation and the preceding generation, especially those who are the journal editors and the society presidents and the department chairs and college deans, they're still largely the old guard. There's enough that they are still the gatekeepers, and they have indoctrinated enough of the upcoming generation as well, so it isn't always a perfect clean delivery of a new generation of thinking. There are those who are open-minded and interested, but they're going to have to bide their time until they have tenure. So that's another, say, five years down the road. Then I think you'll start to see some changes. I think you might see people who take a more open approach to it, or if we have a breakthrough.

So how do you feel mainstream science will accept the existence of Bigfoot? Well, it'll only come with either a corpse or a diagnostically significant piece, a jaw or a molar, or if a new precedent is set of accepting the acknowledgement, the existence of the species on the basis of a novel DNA sequence, a vouchered DNA specimen. That hasn't been done previously. We have used DNA to differentiate, for example, two sibling species or subspecies from the skins and skeletons in museum drawers, but we have the specimens there *a priori*. There is a growing discussion in literature that in the case of rare or endangered species, that perhaps we should advance beyond the Victorian ethic of shoot first and ask the scientific questions afterward and not require the lethal collection of a type specimen and rely instead on DNA collected from hair scat or whatever. So maybe our eDNA study of the coming years will be an interesting test case where we can make the argument that Bigfoot should be acknowledged now based on its novel DNA sequence.

The challenge that still remains, and this is the same challenge we've run into with all previous allegations of alleged Sasquatch DNA examples, is we don't have any Sasquatch DNA identified, or recognized. If we did, or if we do and it hasn't been recognized, it's because they're so closely related to us, even more so than chimps, that we may be talking about a distinction of less than one percent of the DNA sequence of the genome. The chances of detecting that difference with just a very cursory sampling, say of part of a single mitochondrial gene is slim, very unlikely, and so the result is human. It's explained as either contamination through mishandling or just misapprehended human hair samples, for example. The third possibility is we just haven't sequenced enough DNA to be able to conclusively say what it is. Now there is a fourth possibility that involves mitochondrial introgression and was the topic of a fascinating paper recently published in the Relict Hominoid Inquiry (www.isu.edu/rhi). With eDNA we are casting the net more broadly at the potential of getting a sample, but to be able to say it's a non-human sample, you've got to sequence a lot of it.

What's your elevator pitch for open-minded skeptics? That one I struggled with a little bit because as you can tell, I'm not very good at elevator pitches [laughing].

We're only going to the sixth floor [laughing]. Yes, exactly, so it is tough. How do you convey in just a few sentences, something that has taken me over 25 years of engagement to comprehend fully and to appreciate fully? Even an open-minded skeptic, if they're willing to label themselves as a skeptic, invariably has their favorite piece of missing evidence. They'll say, "Well, that's obviously a man in a fur suit, or why haven't we found a body? Where's the physical evidence?". My reaction to that is, "I can offer you an apologist's answer to why I'm not turned off by that missing piece of evidence, but account for me all of the existing evidence. You turn the whole thing kind of upside down on its head and oblige me to justify my belief by explaining a piece of evidence that's not there. But you feel no compunction or obligation to entertain an account for the footprint evidence, for example, or the hair that can't be attributed to any common form of wildlife, or the vocalizations that have been analyzed sonographically and aren't a match to any common wildlife vocalizations. You've got all this body of evidence; it doesn't just disappear because you've got blinders on and you're fixated on this one missing piece".

This happened to me, I presented at a university. That question was asked by the department chairman of the faculty member who was hosting me. This person was a forensic archeologist who was used to dealing with hard evidence, physical and trace evidence of the activities of human populations in the region. "So, where's the body, where's the physical trace?", he asked. I gave him the standard answer about rarity, about taphonomy, the fate of remains of animals that die in nature, etc... Then over dinner, we were chatting and I asked, "By the way, did you find my answer at all satisfying?". He said, "No, I didn't" and then went on to explain why he was so concerned about this missing evidence. I pushed back and I said, "I find it interesting that people with your mindset have your favorite piece of missing evidence, and you're fixated on that to the neglect of everything else, all the other evidence. He says, "Oh, what other evidence are you talking about?". I felt like saying, "Well, weren't you sitting there during my presentation?". I said, "Well, from my perspective, the footprint evidence is overwhelming". His response was, "Oh, well, I'm not an expert on footprints". And I stated, "Well, I am, as a matter of fact. Don't you think that my assessment deserves the same deference that I would give your evaluation of an archaeological site based on your expertise?". It changed the tone of the tenor of the conversation just a little bit. That in a nutshell was the characterization of so many conversations and so many skeptical retorts. I just find it amazing.

I think, Jeff, if someone is not interested in the subject, they don't know what evidence is out there. I think that's what it comes down

to. Right, they don't realize it and assumptions are made and preconceptions imposed on what can or can't be out there.

What advice do you have for new researchers? Obviously, building upon what we just said in being smart and selective in the time and energy you invest in researching reference materials and don't buy into every theory that you read online, as intriguing, or as tempting as it might appear. Often in conversations on this topic with new investigators, they're often interested in understanding where they should go. The rule of thumb that I offer people now is, can a black bear make a living there? There's been published research that has shown clearly a very close correlation between the bio-climatic factors that determine the niche of the black bear with those that presumably determine where the most credible Sasquatch reports are emanating from. If you *Google* "black bear distribution map", there are several good examples that show both historical and current distribution.

It's wise to keep in mind that most encounters are purely by happenstance, it's like winning the lottery. But you're not going to win the lottery if you don't buy a ticket. In fact, you probably should buy a lot of tickets on a regular basis to increase your odds and you still may not win. I quickly learned in my own research that I was spending too much time, energy and money traveling to greener pastures to do my research, when there are reports all around here in eastern Idaho, western Wyoming, Greater Yellowstone ecosystem. It's much more advisable to save your money for equipment and grub and time off work rather than driving across the country. Then as you get to know your area, get familiar, know the ecology, where would an animal go to get food, cover, and water? Those are the three principal things that they must have and then frequent those areas, get familiar with the distribution. Where are these resources found? Where are the berries in the fall? Where are the acorns and the pine nuts? Then get familiar with the other animals and get proficient at capturing those animals on trail cameras or finding signs of those other animals. If you're going out in the woods and you never see a deer or you never see a bear scat, then what are your odds of seeing a Sasquatch? I estimate there's probably two hundred bears for every Sasquatch.

Stake out your areas in as close of proximity as possible so that you can frequent them, so you can buy as many lottery tickets as you can and up your odds of seeing them. Hone your skills, learn the technologies, learn the sign, be able to interpret footprints so you're not distracted and become enamored with a bear track that you found and take offense if I point out to you that. That's always a delicate thing, as a scientist, you are trained early on to submit your findings to peer review and criticism. For the amateur that vague footprint, you know, a glob of plaster, that's their golden ticket to Willy Wonka's chocolate factory. If you threaten to take away their ticket, then they don't feel like they're part of the party anymore or it threatens them personally. They can take considerable umbrage at that kind of hopefully constructive and positive criticism and enlightenment.

How to follow Dr. Jeff Meldrum:
The Relict Hominoid Inquiry: www.isu.edu/rhi

RON MOREHEAD

Best known for the Sierra Sounds, Ron Morehead is a renowned researcher, author, producer, and most importantly, an experiencer. He's been known decades for his ardent research into the Bigfoot phenomenon. The Sierra Sounds are Bigfoot recordings that have been scientifically studied and cannot be explained as any known creature in North America. Ron has documented his personal interactions with these giant beings and produced his story on CD and in his book, *Voices in the Wilderness*. In his most recent book, *The Quantum Bigfoot*, Ron combines decades of experience and research into the realm of quantum physics as it could pertain to Bigfoot. Besides being the Keynote Speaker at numerous conventions, he has been featured on radio programs such as Coast to Coast and on television programs such as BBC TV documentaries and the Learning Channel.

It has been a privilege for me to get to know Ron. He is genuine, open-minded, and personable. I must admit, I am slightly jealous of his captivating and smooth voice, perfect for radio or voice-overs. Some Bigfoot enthusiasts may not want to explore Ron's theories on Quantum Physics as it possibly relates to Bigfoot, although I am fascinated with Ron's willingness to explore and share possible theories on the subject. The Sierra Sounds have intrigued researchers for over 50 years and to this date continue to be one of the most interesting arrangements of likely vocal evidence in existence today.

Where did you grow up? Anchorage, Alaska when I was younger and then Northern California in my high school years. I then entered the military and served for approximately 5 years.

Hobbies outside of BF? I have been a pilot since 1965. I fly to Alaska, South & Central America, and the Eastern United States. I also practice martial arts.

Favorite musicians or bands? I am into all kinds, from country-western to jazz-depends on my environment. I used to test guitars for Fender. I had built my own guitar when I was there. The person who got me the job was a musician friend of my dad's, Bill Carson, and he was the person who invented the Stratocaster.

Most memorable event and where was it? Filming of the Sierra Camp with David Paulides, *Missing 411, The Hunted.*

Favorite place you have visited? Why? Peru. I researched the elongated skulls with Bryan Foerster to see if they could have a connection to the sagittal crest often reported on Bigfoot. I have been to Nepal with Peter Byrne and to Russia and Siberia with Dr. Jeff Meldrum and Dr. John Bindernagel.

What was an embarrassing or funny memory? I couldn't stop laughing when I was quietly told a joke by a friend while in church.

I think that I was kicked out of a couple science clubs in the recent past. They did not agree with my theories. I asked one group, "Classical science is based upon everything being measurable and predictable. Can someone here tell me how far it is from here to the end of the universe?". That stopped correspondence.

What is one thing about your childhood that you look back and say, "What was I thinking?"? When I was about 4 years old, my aunt caught me by the ankles when I was trying to jump out of a 2-story window. I had a towel around my neck like a superhero. I thought I was going to fly. Maybe that's what got me into flying! [laughing]

If you could take one person past or present on an expedition with you, who would it be and why? Joe Hauser (owner of the Montana Vortex). He's a survivalist, a Wildlife Biologist, and an overall great person.

When and why did you get involved in Bigfoot research? I was 29. There were a group of hunters, the Johnson brothers and friends who encountered Bigfoot in their camp in the summer of 1971. There were(sic) 5 hunters who had been hunting there dating back to 1958. It's an 8-mile hike and 8400 ft of elevation. They would hear these creatures around their camp. One of the hunters on one occasion was so shaken, the next morning, he left a note for his friends and left camp. He arrived home and the wives of the other hunters were worried because they had not returned. He asked me if I would travel back up with him to check on his friends because he did not want to go by himself.

When I arrived wobbly-legged from the 8-mile hike in the rugged terrain, the hunters were ok. Fortunately, I was able to ride a horse back out so that trip was much easier.

I then got involved with the group and became a hunter and, by mules, we started packing in supplies to the camp so we wouldn't have to carry so much when we traveled in and out. It was a pristine and remote location by foot. No other people outside of our group would be near the camp. There are a lot of interesting phenomena that go on in that area. Nothing surprises me when we get up there.

Many thought it was a hoax except for us who were experiencing it. We were business people, had careers and doing ok. We didn't need a Bigfoot thing in our lives to screw things up. Warren Johnson, our leader, wrote a 23-page letter to Ivan Sanderson who thought it may also have been a hoax and decided to send it off to Peter Byrne as well. Peter felt the same way and sent it to Alan Berry who was in Reading, California working for a newspaper at the time. Alan Berry started recording in 1972. He was trying to look for a hoax. He was the one who originally fostered the study of the sounds. He first sent the recording to *Syntonic Research Inc.*, the same lab that studied the *Watergate* tapes. They did verify that the sounds were real, although the funding wasn't available to cover their costs for a full study. He then found Dr. R. Lynn Kirlin from the University of Wyoming, to study them unbiasedly. He confirmed that the sounds were authentic without alteration and the sounds were made by a creature with "features corresponding to a larger physical size than man." They also concluded that the tape shows none of the expected signs of being prerecorded or rerecorded at altered speed and hence diminished the probability of a hoax. No one has been able to debunk these sounds for 50 years.

Sierra Audio

If you have had an experience(s), what was the most compelling one? 1974, my friend Bill McDowell and I rode our horses in to bring supplies to the camp. After this experience, we realized these creatures knew who we were and were aware of when we would arrive. The creatures had gone right to the camp. Once it got dark, they started beating on trees and whooping. I also got to see one that night. Typically, they would not interact with us until we were in our shelter, although this time they started prior. We would get glimpses of them through the trees and that was the most memorable experience. I have no doubt that they recognize us. The next morning, we did find evidence, recent tracks and we noticed fresh alder brush had been placed on the roof of our shelter prior to our arrival.

Approximately how many eyewitnesses have you spoken with? I have spent over 50 years talking to people who share their accounts. It is always good, helps you put the dots together. Some of the academia want to put them strictly in the ape camp and just some relic hominid. Based on my research, I feel they are not all the same. Our tracks are different than the tracks found at the Patterson-Gimlin film sight.

What was the most compelling witness account that you heard first-hand? There was a woman from the UK and she was on the coast in California sitting by the water for an hour or so waiting for the sunset. She was intelligent and credible. She saw 2 female Bigfoot and a large male walking on the beach and then walk out into the ocean. The male was going under the water and pulling up seaweed and throwing it back to the females who were chewing on the seaweed and then throwing it over their shoulders. The witness watched this for several minutes and was dumbfounded. Being from the UK, Bigfoot was not on her radar at all. She tried to capture a picture although they heard her fumbling with her camera. As you can guess the witness was extremely nervous. The male Bigfoot screamed and started running out of the water at the witness. The two females stopped him and they were chattering back and forth. The witness confirmed that the chatter sounded similar to what we recorded in the Sierra mountains. They were only 20 feet away from her and she was watching them make the sounds. The witness' profession just happened to be a linguist. She could speak, write, and talk in seven different languages. She said his throat swelled up when he screamed.

She agreed that it is possible they may have more than two vocal cords since they have these unique sounds. This could also explain the whistles that we recorded which project through their vocal cords and not their lips. Gorillas and other primates have large throat sacs so this is not uncommon for primates.

What do you feel Bigfoot are and why? In my opinion, they are not all of the same genomes. Eons ago, beings from the cosmos have altered the DNA of distinct species on earth to facilitate the survival of their species here. Some are relic hominoids, some are not, but all are hybrids. Many have cross-bred with Indigenous people and have become more human-like.

The giants that were with us have a language. Scott Nelson the Cryptolinguist, studied the audio and says they have a complex language structure. According to Dr. Leberman at Brown University, only humans are supposed to have these traits. This tells me that there is a human component to them or there is a Bigfoot component to us.

I think there has been an alien intervention into the genome of varied species, including us. The DNA has been manipulated over eons and all may be hybrid-type beings. I have seen evidence of past aliens, and they are most likely still here on Earth currently. I think they have manipulated the genomes of multiple species to acclimate their species to this planet.

What excites you about this subject? Researching to find out what all these creatures have in common, and how they are so elusive. We all need experiences of our own. I believe in multiple embodiments; our energy which we are made of does not die. It just changes forms. If you are religious, you go to heaven and if you are a physicist you go to another dimension. When we experience things on this planet, we need to respond to them with compassion and love. That's how we progress.

If Bigfoot is a large primate, what environmental conditions would it need to survive? Obviously, it survives in the same conditions that humans do. They need air, water, food, and plenty of room. This planet has all that.

How would you explain the elusiveness of Bigfoot? They are very tuned in to human exploitation and are not supposed to interfere with our decisions. I believe many live underground and possibly alter their mass to energy...going out of our limited perception.

What are your current research goals and how do you go about them? After my personal encounters and over 50 years of researching, my goals are to provide information to others so they can understand what I think I've learned about the species.

What do you feel is your strongest skill/talent that helps you in this field? I am not sure if it is a skill. I had the opportunity to be in the camp and record these things. We let the science support it. Al Berry had a master's in science and would stick with that aspect of what we experienced. I started studying Quantum Physics and while some classical scientists still think that quantum science only exists in the micro world, everything starts at a cellular level and carries on throughout the universe (Dr. Baird, Texas A & M). We live in a three-dimensional environment and see in certain frequencies. There are so many other frequencies that exist. Everything is energy, *Tesla* incorporates energy, frequency, and vibration.

We are all spiritual beings; I am trying to bring science and spirituality back together. My biggest goal is to get researchers to open their minds and accept what is really going on. There are things that we don't *see* with our eyes, our vibrational frequency isn't high enough to get out of this three-dimensional environment. In the past, I would not accept the reports of trackways disappearing. I was not open-minded about it. Peter Byrne used to say [laughing], "Trackways can't just disappear. What did a helicopter come in and pick up the Bigfoot?".

If their density can change from mass to energy, I feel it is possible.

Is there anything that you have changed from how you approached your research in the beginning? I was brought up in a church environment and I knew Biblical scriptures and such. When I had my first experience in the 70s, I went to the Bible to find out where giants came from. Many things did not add up to me, for example, after the *Flood,* how would these giants get here? I feel the bible and scriptures are interpretations controlled by the governing power at the time, who set the narrative.

What do you feel is your biggest challenge that hinders your goals? Getting researchers and others to understand the difference between Newtonian physics and Quantum physics. Quantum physics works throughout the universe. Nobel prize was awarded to Max Planek in 1918, but most still don't (and cannot) comprehend how the universe can have no end.

What is your biggest accomplishment in this field? Recording, experiencing, and relating information about these giants.

Why do you think there is not more evidence or proof? Bigfoot are elusive and smart; they are much more intuitive than many researchers want to give them credit for. I feel academia and the government is aware of these creatures. They don't want the public to know that they exist. Unfortunately, to keep credibility and funding, academia cannot think outside their disciplines with subjects like this.

How do you feel mainstream scientists will publicly accept the existence of Bigfoot/Sasquatch? For the government to tell them it is ok. One of the problems is that we call them "Bigfoot". It is *cartoonish* and prevents some from taking the subject seriously.

What is your elevator pitch for open-minded skeptics? Keep an open mind, just not so open that your brains fall out.

What suggestions do you have for researchers who want to get involved in this field? Start with an open mind and research everything they can with an open mind. Be objective yet understand quantum mechanics.

How to follow Ron Morehead:
Ronmorehead.com
ronm@bigfootsounds.com
Facebook: Ronald J. Morehead

TODD NEISS

Todd Neiss has been an active Bigfoot investigator for almost 30 years. Most of his research is focused in the Pacific Northwest although he conducts research in Oklahoma, Arizona, Alaska, Nebraska, and British Columbia as well. In Nebraska, he collaborates with the Omaha tribe which has a long history with the creature and hopes to set up protection for the species through their collective efforts.

Todd did not grow up with the fascination of Bigfoot or other cryptids like others in this field, he had an encounter while performing military training in 1993 and it changed the course of his life forever. He along with 3 others in his platoon were eyewitnesses to 3 mystery creatures that day on the Oregon Coast. I reached out to Todd and he was eager to help me with this project. Also, he generously connected me with pioneer researcher, Peter Byrne.

During our interview, one of the topics we discussed was Native American history with Bigfoot and specific literature that Todd has found informative over the years. Later that week, the book *Bigfoot/Sasquatch Resurgence of Native American Indian Legends* by Leon Pfaller showed up at my door, courtesy of Todd. The book features historical accounts of the hairy creature, all with unique names from 99 tribes in North America.

Todd and his wife Diane, who is also a long-time researcher founded the American Primate Conservancy. They created *Beachfoot* in 2008, which is an annual invite-only conference for researchers of the Bigfoot Phenomena. In addition, since 1993 he has organized and led expeditions and investigations collaborating with individuals such as Larry Lund, Peter Byrne, Thomas Steenberg, Ron Morehead, Joe Beelart, Dr. Richard Greenwell, and others. Todd has been featured in books, television, radio, podcasts, and news media. Since first sharing his experience at a Bigfoot symposium, he has spoken at numerous events and interviewed hundreds of witnesses.

Where did you grow up? I was born in the Pacific Northwest, in the Portland suburbs.

What is your favorite type of music or musicians? I was mostly into classic rock for many years until I attended a friend of mine's country-themed wedding in California. I was the best man so I had to go out and get a Stetson and the bolo tie and all that good stuff. Anyway, I spent a couple of weeks with them and I really got into country music.

What is something that most people do not know about you? I raised three non-biological sons, on my own, for 10 years. I'm "Dad"

What was an embarrassing or funny memory while researching or any moment in your past? While filming the FOX TV series *Encounters* in July of 1994, I was beta testing a unique camera system I had invented. The director, Chip, asked me where I was going to emplace the system. I told him I would set it up in a nearby rock quarry, to which he insisted was not appropriate nor camera-worthy. I explained that Bigfoot actually utilize quarries as squirrels hide under the rocks for protection as well as heat. In fact, my sighting of three creatures the year before was in a rock quarry. He laughed at the thought. Seconds after he started laughing, a squirrel suddenly appeared out of nowhere and ran right over his feet! That put an end to his mockery.

If you could take one person past or present on an expedition with you, who would it be and why? Dr. John Bindernagel. We met in 1996 at a conference in Harrison Hot Springs, BC. We became friends and unfortunately, I never got to go into the field with him.

How did you get involved in Bigfoot research? In 1993 I was a sergeant in the Army's 1249th Combat Engineer Battalion. Our mission was to conduct training on private timberland near Saddle Mountain. We were executing demolitions (explosives) operations at three rock quarries. Each site had a unique battle scenario to carry out.

Being a squad leader, I had the privilege of having my own Hummer, I assumed a position behind the driver's seat and, as we were descending the narrow winding road down towards the staging area, I had the opportunity to enjoy the scenery. As an avid hunter, it is just second nature to me to spot wildlife. As it was a rare sunny day in April, I had my window unzipped for a better view. Rounding a corner, I had an unobstructed view of the rock quarry where we had done our second blast less than an hour earlier. Standing right out in the open, in the middle of the gravel pit, were three, jet-black, bipedal creatures. They stood in line (shoulder to shoulder) staring directly at our convoy as we descended the hillside across from them. I could not make out facial features or gender, but there was no doubt what I was looking at were not humans. Had these creatures been standing in front of a backdrop of trees, I most likely would not have seen them at all. But in this case, there stood three dark black figures contrasted against a light grey cliff of basalt on a bright sunny day.

In the middle stood, what I assumed to be, the alpha male of the group as it towered a full head above the two creatures that flanked it. I would estimate it to have stood approximately nine feet high, with the flanking creatures approaching seven feet in height. Their silhouette was unique in that their heads sat directly on their shoulders with no visible necks. They all displayed broad, square shoulders and barreled chests which tapered down to a svelte waistline, unlike the creature seen in the Patterson-Gimlin film of 1967 (for the record, I am of the impression that the PG creature was either pregnant or had recently been so; accounting for her girth). The arms of these beings hung well past their knees. In the case of the two flanking creatures, they were exhibiting a swaying motion (rocking side-to-side) as the larger creature stood as still as a statue. Bear in mind that, all the while I was staring at the creatures, we were bounding down a dirt road with the occasional hedge of blackberry and Scotch bloom obscuring my view. I had approximately 25 seconds of viewing time.

If they exist, what do you feel Bigfoot/Sasquatch are and why? Relic Hominid. Native Americans/First Nations people have spent over 19,000 years living with them. They refer to them as a tribe: the "First Ones" or the "Old Ones."

What type of research are you currently involved in? Bringing awareness of their existence through science and education. The goal of the American Primate Conservancy is to establish international protections for this species. I am currently collaborating with several tribes to do just that. I recently drafted a regulation for the Omaha tribe entitled "The Omaha Nation C'Tonga [Bigfoot] Protection Act." I believe it will catch fire with other tribes and ultimately make it more challenging for the US Department of the Interior to ignore.

What do you feel is your strongest skill or talent that helps you in your research? To be honest with you my personal experience. I'll expound on that; say you have two people. One who believes these things exist, they want these things to exist in their hearts, they know they exist, they wish they exist, but they've never seen them. This is nothing against them. On the other hand, you have somebody who's seen them, who's had that experience, who now has no option but to admit that they exist. You can't unsee something like that. That contrast to me makes a significant difference in your approach to the subject. Going in knowing, really gives you an advantage, a distinct perspective. I don't wish, think, or hope they exist; I know they exist so now I can take that off my list and just focus on finding the evidence and not be distracted by the whole idea of wanting or hoping or guessing.

Does that help with interviewing witnesses? It plays to a great advantage when it comes to interviewing eyewitnesses. When you let that eyewitness know that you've been there, seen that and share that experience, they open up like they never would to anybody else because now they don't have to in the back of their mind, wonder whether this person takes them seriously or whether they think of them as a crackpot. It's kind of like belonging to a fraternity or a club or something. It's difficult to go through your life with that sort of experience and not be able to compare notes with somebody else who has.

Is there an eyewitness report that stands out to you? One that jumps out at me was one that I investigated on the Roaring River, a tributary to the Clackamas River. This gentleman had gone in about a mile upstream, very rough hiking, no real trail to speak of since the '96 floods washed them out. He made his way about a mile up the creek because he realized that there was this lava flow that crossed the creek, and when the summer comes around and the water table drops, it exposes much of that lava. Embedded in that lava is what he called common opal.

He and his full-grown Australian shepherd hiked in, and down to the riverbank. They were chipping away at this common opal. Before too long, his dog started getting upset, it was growling, and since he was down on the riverbank, he couldn't see what the dog was looking at.

He decided to investigate and looked in the same direction that his dog was staring and growling. He said, "There was a gorilla standing there". He never used the word Bigfoot, he said there was a gorilla standing up on two legs glaring at my dog and grunting with every breath it exhaled, and it did not look happy. They were in this little standoff; we would later measure the distance at 47 feet. As soon as he saw that, he pushed the dog in front of him and ran as fast as they could. Quite unusually, this particular animal decided to chase after them, which rarely happens. He said he had no doubt that if this thing really wanted to overtake them, it wouldn't have any problem doing so. In hindsight, he realized that they were just being escorted out of that area, but that that stood out as an unusual sighting.

The most bizarre thing was on our way back out during my investigation, taking the same route, we ran into this wall of smell, if you will, that reeked very strongly of rotten meat or roadkill. It wasn't there 30 minutes earlier so something within 30 minutes had been at that spot and had somehow left that odor. It was very concentrated; you could step into it and step out of it. You would think something that pungent, that rank would drift quite a ways, but it did not. It was very centralized, localized to that one spot. We could never find the source of it. I have no doubt just based on all the interviews I've done in the past that there was a Bigfoot there, probably heard us coming and in and beat feet, but left us his calling card.

What do you think about eye shine? I was invited to be a guest on the Omaha Indian reservation, where I saw this phenomenon not just once, but five nights in a row and at two separate locations. It seemed to be very spontaneous, it seemed to possibly react to some of these calls that my friend made in his native language. It wouldn't take more than 30 seconds after he started trying to contact them verbally that what appeared to be glowing eyes with no ambient light whatsoever would seem to be not far from us, 80 to 100 feet away. At times there were two, even three at a time and they would move, they would blink. Occasionally they would grow dimmer or even go out and then come back. I assumed that they were probably the brightest when they were staring directly at you and if they were looking left or right, they may be dimmer. I saw this with my own eyes, I can't unsee it. I fully intend to investigate that phenomenon and find out if there are any natural explanations for that sort of bioluminescent behavior.

What are the goals of the American Primate Conservancy? To find proper evidence, hopefully, something that yields DNA that can't be dismissed and to get them officially recognized and hopefully protected someday.

How do you feel or is that how you feel mainstream scientists will finally accept the existence of Bigfoot? Well DNA cannot be argued against, and that's what everybody is kind of waiting for. We've had hair samples, which are not very good at giving DNA unless you were to get the follicle itself. It's going to take somebody finding a carcass or something that has at least enough genetic material on it to be analyzed.

Beachfoot 2014 - Photo courtesy of Todd Neiss.

What would you recommend for a new researcher? Step one, do your own research, that'll keep you busy for a while. Step two, get out in the woods, it's kind of like the lottery you can't win if you don't play. You're not going to have an encounter in some urban environment, you need to research, find out what sort of habitat that they prefer, where they're at certain times a year and ideally identify a specific location that's had recent activity and then get out there. I would not recommend anybody go alone; find somebody you trust. Step three, if you feel the need, join a group, and find a Bigfoot community that you feel is responsible and realistic. Its buyer beware when it comes to groups because some of them have certain agendas and you just got to kind of weed them out.

How to follow Todd Neiss:
www.Americanprimate.org

RICHARD NOLL

Rick Noll (he prefers being called Rick) has been researching the Sasquatch phenomena since 1969. He has worked with René Dahinden, John Green, Dr. Grover Krantz, Peter Byrne, Dr. Jeff Meldrum, Dr. John Bindernagel, Dr. Daris Swindler, Dr. Briggs Hall, Dr. Bob Walls, Dr. Henner Fahrenbach, and numerous other academic and vocational experts. He was an investigator for the Bigfoot Field Researchers Organization (BFRO) but has since returned to independent research after the Skookum Expedition. On that expedition, Rick, Derrick Randles, and the late Dr. Leroy Fish discovered and cast the Skookum impression which is one of the most unique and fascinating findings in the history of the subject. Researchers involved in the subject can see an impression of a large primate-like body resting on the soft ground while eating fruit. Along with the impression, chewed fruit remains with tooth impressions and potential hair samples were recovered.

Rick has displayed the cast of this impression at the Museum of Man in San Antonio, Tx, and to Dr. George Schaller and Dr. Jane Goodall. Dr. Schaller commented that in all the ungulate beds he has discovered and studied he has never seen something like this. But on the other hand, an anonymous scientist that works with smaller fossilized tracings (worms), as an aside to his oil industry employment, after having seen the museum copies presented at the Museum of Man, stated he thought it was from a common Elk, comparing his picture of the display to a field book showing North American animals.

Rick has worked on two TV series with White Wolf productions and multiple TV networks' special programs. His material has been used in several productions including books, magazines, and videos. He has also presented at various venues. He was hired by John Green to digitize every frame from the Patterson-Gimlin 16mm film using a modern DSLR and high-quality optical microscope.

There are few individuals that have such an extensive Bigfoot history as Rick. He could author his own book on his relationships and exploits with the early researchers. His attention to detail and innovative approaches to research sometimes get overlooked, although he is a pioneer in his own right. The impact of Skookum Cast evidence continues to inspire Bigfoot researchers and is compelling to scientists who are skeptical of the possible existence of the elusive creatures.

What is your favorite place you have visited and why? The Galapagos Islands. The similarity of life between the islands, both above and below the water line, accentuates their differences and yet links them in the evolutionary demands of survival. It illustrated to me the potential for all life to co-evolve when presented with the minutest of differences. From one island to another, the same species of birds evolved different task-specific beaks for the types of resources found on them. Species isolation is very evident what with it being so close to one another. Underwater on one side of an island is the Humboldt Current with happily swimming penguins, yet 200 feet away are warm water sea turtles, manta rays and tropical fish. These turtles illustrate overpopulation when they hatch in the sand and turn to the nearby ocean but with the vast majority picked off by Frigate birds or sharks, only one in a hundred make it. Of course, the history there is very inspiring, sad, and embarrassing. The observational skills Darwin had, the plight of ancient seafarers placed on the islands releasing invasive species and love triangle murder mysteries all make one wonder about human influences and what is to come when Exoplanets become explorable.

What is something that most people do not know about you? I was used as a pawn in a bitter feud between the *Four Horsemen* investigating Sasquatch. They would attempt to get information from me about each other's activities. I have many interesting stories about them. It started when I was investigating a Bigfoot trackway, discovered in North Bend, WA and both Peter Byrne and John Green tried to interview the discoverer of the tracks at his home, at the same time. Rene Dahinden had also sent some investigators to the site ruffling even more feathers at the actual track site. They ruined quite a few tracks from the site and finally gave up trying to cast them. Back then, even cameras were rare for investigators.

It continued when everyone who had anything to do with the subject attended the famous UBC conference in BC in 1978 (Manlike Monsters on Trial), the one where Dr. Krantz started his presentation on the ballistic characteristics of the ammo he carries in the field, sparking the "To Kill or Not to Kill" debate. Dr. Barbara Wasson wrote a Bigfoot book (Sasquatch Apparitions, 1979) and tried to anonymously chronicle the cloak and dagger events, as a side liner, at the conference. I got tasked by Rene to investigate the White film (along with Tony Healy and Bob Walls) and sat in on the first listenings with Peter of the Sierra Sounds. I first met and talked with the famous author Peter Matthiessen at the conference, Peter had already traveled in pursuit of the Snow Leopard with Dr, George

Schaller and would later relate the search in one of his most acclaimed books.

What is a bucket list item that you would like to accomplish in the next 6 months? Establish a point of departure for a book and documentary film on a famous past Sasquatch encounter. One that hasn't really been in the mainstream Bigfootery.

What was an embarrassing or funny memory while researching in your past? Peter Byrne had a trailer that he used as a headquarters for his *Bigfoot Research Project*. Once he organized a meet and greet with wine and hors d'oeuvres and he had all these fancy things going on. Some politicians attended, I assume he was hoping to get some publicity and maybe some state funds, I don't know what he was thinking of. He'd already gotten a trailer, the land for it, and grants for fuel and his living expenses. He also acquired vehicles, a plane, and camera equipment. Anyway, I was serving everybody wine, which was my job. In the middle of the event, we received a phone call, and I thought, "Oh, yeah, just perfect. Everybody's here, all these people you're trying to impress". It was a witness who said they had a Bigfoot on their property. So, Peter pulls all of us in and it's *hush-hush* as we all crowded near the phone to listen to the conversation, it was funny. In the morning, Peter and I were off to the Cascades in Oregon to investigate that report.

Acting as a guide for a group out of Olympia, WA, searching for Bigfoot evidence near Mt. Saint Helens, we first stopped and talked with the famous Harry Truman (later killed by staying during the mountain's eruption 5 years later) and if he had any knowledge on the subject. He told us there was no such thing and I told him he might just be surprised. He called us fools and told us to leave his property and said we should take the shortcut through some hedges in front of his place. We went in through an opening to return to our cars but soon found out that it was a maze. Took us a while to get out of that thicket, I still imagine the crusty old guy sitting on his porch laughing to himself.

If you could take one person past or present on an expedition with you, who would it be? For the most part, a lot of the scientists I have worked with have been disappointing in living up to the hype, created in my mind. There are a few that did live up to it and have even gone beyond. It would have to be one of them for sure. They would need to be able to dedicate a significant amount of time to the expedition, not just two or three weeks. I have two people in mind, which have proficiently level heads on their shoulders, and are not afraid to rub things the wrong way.

I would have loved to have gone out with Darwin but I don't know how many allusions to religion I stand in the field.

When and why did you get involved in Bigfoot research? My brother and I spent the summer of 1969 with my cousin Harry down in Redding, California. My cousin was building bridges and was constructing one outside of Happy Camp on the Klamath River. My brother, another cousin and I would spend time there while Harry was working. One day my brother and cousin were fishing and got swept underneath the water, it was frightening. I grabbed the axe we had at camp and fortunately, was able to pull them out, they could have died, it freaked me out and was upsetting. To keep my mind off the incident, my cousin Harry put me to work with his crew and started telling me Bigfoot stories to distract me.

That sparked my interest and when I returned home, I purchased John Green's books at a local grocery store. A brief time later, I called John Green and asked how I could get more involved. He connected me with Dick Grover who was closer to me and was researching the subject. Once, I traveled with him to Olympia, which is the state capital of Washington. We had track casts and flyers that we handed out. We were trying to convince legislature to make Sasquatch the state animal, it was weird. I then entered the military. After my time in the service, I went on to Green River Community College to study anthropology and forestry. While I was attending college, David Smith (the Harry Truman story) had formed an expedition at Mount

St. Helens. I hiked that area my entire life so I contacted David and asked if I could join the expedition and offered to guide them.

Later as a classmate, David Smith and I formed a Bigfoot club at the school. Through word of mouth and friends in college, I would get reports. Eventually John Green, Grover Krantz, René Dahinden, and Jon-Eric Beckjord started reaching out to us on a regular basis to investigate reports in the field. Eric also went to the UBC conference, and got kicked out of the presentations because he couldn't stop himself from heckling the presenters so he sat out in the hallway entrance with a display... a bent car antenna that he claimed was held in the mouth of an angry Bigfoot. This was also the time of the Puyallup tapes and Mark Pittenger's encounters

I started going to Bigfoot conventions, like the one that used to occur regularly in Harrison Hot Springs. New researchers were starting to attend which was exciting. Dr. Jeff Meldrum took interest and he would bring along friends and then other scientists would also show up. Soon many people with credentials were attending the conferences. I would have to talk René down from getting really upset, preventing him from causing a scene. He didn't like the intrusion. He believed everyone wanted his information, but they didn't want to give him any credit. He felt really slighted about that. He only authored one book. I don't think he could write one himself so he had someone else ghostwrite it for him, although it really wasn't a ghostwriter because René's name was still on it. Truthfully, it was a pretty funny book at the time compared to all the others. John Green's book was written like it came straight out of a newspaper.

I noticed that with all the Sasquatch activity I was involved with 80% of it came to naught because that's what it was. It was the 20% left over that kept me going... finding those damn tracks out in the woods. What was making them? Why haven't I seen one of these creatures?

What else can you share about the early researcher's personalities?
Peter Byrne was, I guess still is, in *The Explorers Club;* I think he

wanted to find something that no one else did or something [laughing]. He did blaze the ascent route for the 1st Everest summit. I think that is what got him in the exclusive club. René Dahinden was just a down-to-earth type of person. He had some strange concepts, although I came up with a lot of weird ideas, as well. Later in life, I worked for Leica and told him that I visited his hometown in Switzerland, Lucerne. He said "So Vhat!" I was one of the first to suggest using mountain bikes while researching. I would ride around in the woods and on these dirt roads at night, with a camera on my chest. If something popped out in the woods, I would take a picture of it. I captured pictures of bears while on the bike. René loved the idea but was in his late fifties at that time. He really wanted to try using a mountain bike like mine. I thought that this may be a little too dangerous for him. I could see being the cause of an accident and forever being labeled that. So, instead of getting one, I steered him to going the dirt bike route and he went out in the woods on that and then said it was too hard and went with a 4-wheel ATV. Probably scared away all the wildlife there was with the loud exhaust.

Since René was not a US citizen, he could not enter with his rifle. Somehow though, he bought one in the United States and stored it in a locker. He would pick it up whenever he came down, and then go out in the woods with it. Peter Byrne said that he was peaceful and he would never shoot a Bigfoot. Although underneath his International's back seat was a 7mm Weatherby. When asked about it he said it was for some unsavory characters he had met in the field in the past. I let it drop but believe he was talking about René there. This time period had these researchers acting like gold miners.

Grover did some strange things, too. He had one paper out that really surprised me. He challenged his students to figure out why the *early man* had such a protruding brow ridge. After his students wrote up their papers, he said, "Now I'm going to show you what it was for." He then made a fake brow ridge appliance, attached it to his own brow, wore it all summer, growing his hair long and then wrote a paper on it. It was bizarre. His theory was that it kept his uncut hair out of his eyes, a survival advantage. He was a weirdo about stuff like that. But he was a funny guy. In fact, if I had to equate him with someone now, I would say he reminds me of Donald Trump. Dr. Krantz once touted that he was the equivalent of Leonardo Davinci when it comes to Sasquatch research. Then again, he would often ask me if I wanted to know how Hitler could have won the war. That wasn't funny.

If they exist, what do you feel Bigfoot/Sasquatch are and why? Best guess, an evolved Gigantopithecus. I am not sure if we really know what all MegaFauna existed here in America, in the past. I suppose the possibility of a remnant limited population of an upright short-faced bear might be possible. You know about Saber-Toothed cats? They are not really cats. They are marsupials. Co-evolved with other marsupials in Australia. They carried live young in a pouch. So maybe the Sasquatch is a co-evolved primate, reminiscent of those found elsewhere but without trace evidence in the form of their fossils being discovered yet.

If Bigfoot is a large primate, what environmental conditions would it need to survive? Using their observed activities, I do not think they require a high-energy diet like a chimp, more like one similar to a gorilla. Where the reports come from, indicates to me that they are large and can regulate their body temperature in cold environments. They are intelligent but not as much as many would romanticize them to be. No reports of fire building, tool construction, or usage. No one has seen them make a nest or a shelter, they seem to be entirely a solitary intelligent opportunistic animals. I also really do not think anybody has habituated one or a family of them yet. The reports I have investigated where this was alluded to also had a romantic aspect to them. Between family members, lack of it between family members, wishful thinking, escapism, hopeful in drawing attention... human emotional conditions that get explained by the incredible. Someone once said that to believe Bigfoot can communicate using ESP with humans would make world news about ESP, but sideline the discovery of Bigfoot.

How would you explain the elusiveness of Bigfoot? Minimum population (in WA maybe only as many as 50 could exist without a massive amount of observable environmental impact), solitary lifestyle (adult males range alone, females alone to but move to a lesser degree in range for mating and child raising), large territorial range (maybe larger than 200 sq mi), very protected and hidden home range (no human impact), in death they may seek out hidden protected holes which could become their grave (hiding in hopes of recovery and not predated upon at their weak moment), most if not all human encounters resulting in their misreading of our behavior. There are 130K deer in Washington, we see deer all the time. There are 35K black bears in the state, and seeing one is rare. There may be 15K cougar, even rarer. 300 Wolverine in the lower 48 states, super rare and this is where we are with Sasquatch reports that amount to something.

Rick Noll and David Smith sent to the Olympic Peninsula by Peter Byrne to investigate a report of a Sasquatch carrying a dead sea lion across the road. Photo courtesy of Rick Noll.

What type of research are you currently involved in? Researching past hot spots and comparing them to more recent ones. What makes a hot spot? How do the demographics and behavior of humans play into encounters? What else could the Sasquatch be? Are there holes in the environment that they could exist in or slip through? I think it important to document both the observed behavior in encounters as well as the researcher's behavior and any other observable circumstance that could influence the behavior of the creature during these encounters. Maybe a significant number of encounters are by those wearing yellow raincoats. Maybe tracks are best found after heavy rains or behind forest road cut bank vegetation. Maybe loud noises pique curiosity in attracting them closer but are only observed by the quieter members of the group of people in the area.

What researchers past or present have influenced your work? I have gotten the most influence from researchers in more traditional fauna studies. They tend to distill their very specific work down to more universal nuggets of knowledge that can be applied to many other animals and their behaviors. But if the question is specific to the Sasquatch, all of them but the wackos and I think they know who they are.

What events led up to the discovery in Skookum Meadows? In 2000, I was doing metrology work for Boeing. Computers were not that prevalent and I felt like I needed to have a computer at home. So, I convinced my wife and we purchased one. Most Bigfoot communication back then was on bulletin boards and chat rooms. I connected with Jeff Lemley, who was with the BFRO and we made time to meet. I eventually joined an expedition at Mt Saint Helens in south-central Washington state. René Dahinden warned me to watch out for the BFRO because the internet was kind of strange and the California people were the same way [laughing]. I wasn't too impressed, but they did have some interesting things going on. They would play loud supposed calls of Bigfoot, maybe at 500watts, and have people stationed at various locations some distance away to listen for a response. They then went in chase.

Later, another expedition started forming and Matt Moneymaker wanted me to join it and I agreed. I told him though, "If I make a cast, it's mine and anything I photograph is mine". He said, "Okay, no problem." We arrived and Moneymaker showed up, and then a Discovery Channel film crew from Australia showed up. They interviewed everybody except me.

One of my theories at the time, as well as today, is, for a Bigfoot to thrive, it's probably not a high-energy diet that keeps it going, like with fruit. Fruit more so than a lot of other things like meat. Like chimpanzees and gorillas. Gorillas are not as high-energy as chimps, and they eat a considerable amount of fruit and vegetation. They do eat meat, but it's far and few between. So, I don't think Bigfoot are out there killing calves of elk and deer on a regular basis. That can be dangerous, they could get injured. If they're not that plentiful of a population, they could easily be hurt and die out and the entire population could be at risk of becoming extinct. Based on that theory, I have never used bait, plus there are laws governing doing so. Well, there were a few arguments amongst the expedition members to bait, to strategically place fruit in locations in our expedition area. Reluctantly I agreed finally and participated one night near the end of the expedition, driving around, cutting up fruit on the road, being filmed with a thermal camera, looking for latent warm tracks on the ground. We found a few tracks that looked promising and I got excited and went further and further out from our base camp in my vehicle, placing fruit and having the base camp call blast... till my truck stopped running. The alternator died and I had to be rescued by team members and the battery jumped. When I got back, others took up the effort and placed fruit right where the Skookum impression was found the following day.

While back in base camp, with the others continuing the baiting process, Matt and I heard what sounded like a large animal come close to our campfire and try and mimic our conversation with guttural sounds. Matt was of the opinion it was just a team member sleeping in a nearby tent. I didn't think so. I was a sonar tech in the military and pride myself on having very sensitive and accurate hearing. The sounds were not coming from the nearby tent. Searching for tracks the next day where those sounds came from produced nothing, but a ways away I found a few possible soft-footed tracks that were not there before. I know because that is one of the things I did was to check as much of the immediate area as possible before we commenced "call blasting".

The next morning, we went out in teams to investigate the areas where we placed the fruit the night before. At the first location and to our surprise, all the fruit was gone and there was a strange yet compelling track left next to it, up on a cut bank. I photographed it and then we examined another area where Derrick Randles and wildlife ecologist Dr. Leroy Fish placed a large pile of fruit. We instantly noticed even more strange markings, just really bizarre looking. At first, they didn't even appear to be organic in appearance, more from some mechanical object, like a car's rear differential. It reminded me of an impression of a vehicle with flat tires stuck in the mud. We couldn't figure out what we were looking at and it finally dawned on us that it resembled someone who had sat and rested on the ground. We could see a butt imprint and where it rolled to the side and seemed to reach out and over to where the fruit had been placed. There were scrapes that looked like fingers, from missed attempts at grabbing the fruit (I later filmed gorillas being feed apples at the zoo doing exactly the same thing in the exact body position.) The fruit was eaten and pieces had fallen into the depression where whatever sat there had been sitting. We could see what looked like bites or teeth marks on what was left on several pieces of the fruit.

We cast the entire impression and transported it back to Derek's home. It was examined by Dr. Henner Fehrenbach, Dr. Esteban Sarmiento, Dr. Grover Krantz, and Dr. Jeff Meldrum. They picked out numerous hairs. Fehrenbach found, I think, two or three of them that were of primate origin and that he couldn't identify. The medulla was missing and it was clear that the tips of the hair had never been cut like human hair would from scissors and such. They were broken at the ends.

Skookum Cast

What did you use to cast it? *Hydrocal B11*, it's a higher density type of plaster. It can flow into very small areas and harden there with a lot of strength, retaining accurate representations of surface features as small as 0.010". Once it's set up, the surface on it, it's almost like glass. Any type of deformation that you add to the cast after it has dried and cured or any type of tool marks that you put on it, are definitely visible. On this cast, we labeled the "Skookum Cast" (a Native American word equating to powerful, like with the Skookum Chuck River, the powerful river), you can see areas where the skin was on the ground and fingerprint dermal ridges and hair in other areas. One scientist actually pointed out where the genitals were, it really is fascinating to look at. For display purposes, John Green had a set of copies made to show people when they could go to the actual cast. The way it was made reduced the resolution of the surface to 0.050", smoothing over many details. The copies are lightweight and can hang on a wall, they are also subject to thermal expansion, more so than the original plaster we made the cast with.

What suggestions do you have for researchers who want to get involved in this subject? Go camping, go places where most humans don't travel. Keep your wits about you, don't go running off in the middle of the night chasing after the sound of a breaking branch and get your eye poked out. These animals, should they really exist, are more likely to be on the edges of civilization where people rarely interact with the natural environment. Utilize the evidence in John Green's database. I have entered over 200 encounters into the BFRO database from Washington State, so you can check out the BFRO Flats database. You can do what I do. I use electric bikes now; they are silent and quick. Doug Hajicek helped me build one from easily purchased parts. I can travel about 70 miles in pedal-assist mode.

Most animals stay in the shade, they're underneath things and don't come out during the day. If you go out at night, you can't see anything. It's usually during the fall and the winter that are productive for me.

Come up with a theory and then do your research, be proactive and don't ambulance chase. Record your positive and negative data.

Learn to use a camera and interpret the results accurately and truthfully. Learn to navigate with a map and compass. Research the past work and don't assume you are discovering something unique right off the bat, it can become a bias.

Get out of your vehicle. Walk into the forest further than 100 feet. Be prepared to record in some way what you find. And for God's sake stay safe.

DANIEL PEREZ

Daniel Perez has been investigating Bigfoot for over 40 years and has researched with many of the historical and impactful individuals on the subject. Impressively, he may be the unofficial historian on Bigfoot as he has the largest physical files on Bigfoot in the world. Daniel started his newsletter *the Bigfoot Times* in 1998 and it may be the only mailed monthly circulation in print today. Over 850 readers around the world are kept up to date on topic-related news and receive *the Bigfoot Times* printed on its iconic bright yellow paper. Daniel recently was recognized as the 2021 Cryptozoologist of the Year, awarded by the International Cryptozoology Museum.

For the brief time I have personally known Daniel, I find him to be extremely genuine, kind, and helpful and I enjoy his sense of humor. He has answered many of my topic-related questions, including some that were not for the book. An avid track and field athlete and aficionado, I certainly know who I am choosing for my 4x400m relay team! You may not find another researcher more thorough, concise, prompt, and informative about Bigfoot, which are some of the many reasons he has been impactful to many enthusiasts. If a researcher needs historical information on Bigfoot, Daniel is typically the first place they go.

Where did you grow up? Norwalk, California – born and raised

Hobbies outside of BF? I used to be an avid long-distance runner, but since I am older now I really cannot run as much as I used to, so now I am a track and field enthusiast. I go to track meets and watch the big ones. I would have been in Tokyo for the Olympics if it had not been for the pandemic. Most of my travels are related to Bigfoot or track and field.

Favorite musicians or bands? I really like Carly Simon and just recently I started watching more of her music on YouTube. I found out that she is the daughter of the co-founder of the publishing company *Simon and Schuster.* So, because of my love of books, I learned about her family and by sheer coincidence, she just happens to be a singer that I gravitated to!

Favorite Place you have visited? Yosemite, California. There is NO place on the planet quite like this place.

What was an embarrassing or funny memory? Giving a talk on Bigfoot with my fly open. After the presentation, it was pointed out to me my zipper was down.

If you could take one person past or present on an expedition with you, who would it be? René Dahinden. Never really been out in the field with him in the woods.

When and why did you get involved in Bigfoot research? That is easy, prior to a movie I saw at the theatre at 10 years old, *The Legend of Boggy Creek*, approximately in 1973, I knew nothing about the subject matter. That was the flame that ignited everything! It took several months for me to tie in the creature from the movie to Bigfoot and realize "Oh so that's what they are talking about". Decades later, I am still involved, the interest just grew over time. Although I do regulate it well, it is like an addiction; you get started and you want to feed that addiction as time goes on.

If you have had an experience(s), what are the most compelling one(s)? I think seeing evidence for myself, it was 1979 was I was still in high school. There was activity: sightings and footprints being reported in Hemet, California. My acquaintance, Doug Trapp, who was interested in the topic called me and said, "Why don't we go see Dick and June Putnum and see if there are any recent news on sightings and footprints". We later arrived at their house and no one was home. I suggested that we look around, it was a nice day a little bit chilly although it rained recently. We were traveling on a rutted dirt road so I suggested that we stop the car so we don't get it stuck and explore by foot. We eventually arrived at an area where there was a small foot creek and a scattering of trees. We looked down at approximately the same time and said, "Holy Crap, what is that"? There were footprints, because of the rain, they were badly deteriorated. One of them though, we could clearly see the heel print at the water's edge. It was incredibly big; I have a 10-plus inch foot and this thing was so much bigger than mine. Doug decided to make a plaster of casting although the bag he had in the car had already absorbed some moisture so it did not hold up and was a failed attempt to make a casting. I did get photos though and that was the first time I saw anything in the field and became a first-hand witness, it got me into the mindset of, "oh my god, they really do exist".

What was the most compelling witness account that you heard first-hand? In Ohio, 1996, I interviewed Patrick Poling from West Mansfield who had a sighting while he was farming his fields in 1980. Originally, he was very reluctant to share his account and when I spoke to him on the phone and did not let him hang up and convinced him. He wasn't looking for any sort of publicity and finally agreed to a face-to-face interview. He had a broad daylight sighting, and Patrick was on his tractor. He had always kept a shotgun on the tractor, he saw this "thing" that butted up to the edge of the woods and the farm field. He was not sure what it was although was curious enough to get closer with his tractor, and he felt safe since he had his shotgun. As Patrick got closer the creature retreated into the woods. Later tracks were found. I cannot say enough about his credibility as a witness. He stated, "this is what I saw and I have never seen anything like this, and there it was. If you want to call it a Bigfoot then so be it".

I have interviewed Bob Gimlin numerous times and of course, his eyewitness testimony is accompanied by Roger Patterson's historic Bluff Creek film from 1967. He too comes across with stellar down-to-earth credibility, there is no other way to slice it in my view.

What do you feel Bigfoot/Sasquatch are and why? Primates, like us. To be more specific would be guessing, when, in fact, NO ONE knows. Quote from René Dahinden, "We can sort things out after one is collected, but until then it is a guessing game".

What excites you about this subject? That it is an ongoing mystery. I am not a hunter so I don't know much about it. I do suspect though hunters may not travel into areas of suspected Bigfoot. If they did, this issue should have been settled a long time ago. I also wonder if it is the fact that the population of the species is so low or that hunters do not overlap onto Bigfoot territory. I am fascinated that the mystery has gone on into this century and the witnesses and researchers are fascinating as well.

What inspired you to create the Bigfoot Times? My inspiration for creating the Bigfoot Times was the other newsletters out there at the time, which I felt could be better in research and writing. Think of it like a car. You can see that it is dusty. You pull out the hose and give it a rinse. That takes the dust off, but it is still dirty. If you can grab a bucket of soap and a cloth and give it a good wash, it is even cleaner. If you give the car a good wash with soap and water and wax and polish it, now you have really put some elbow grease into the matter.

That is how I like to see my Bigfoot Times, going into the 24th year in January 2022.

When I started publishing there were many mailed-out newsletters. For example, I believe two were coming from Ohio, one being Don Keating's newsletter. One from California, the Bigfoot Co-op and in Oregon, Ray Crowe was publishing as well. I just thought all of them, which I subscribed to at the time, needed more elbow grease.

That is the story in a nutshell. I am so proud of the newsletter. To my knowledge, it is the only mailed-out Bigfoot newsletter in the world.

What impact has social media had on Bigfoot Research? The aspect of social media and the internet is interesting and how sometimes individuals are marketing and merchandising themselves, instead of doing real research and investigations, they are more focused on" likes" and popularity to confirm how good they are. I don't know if that is an accurate gauge of whether you are a good researcher or investigator. The dynamic of the people involved in Bigfoot research has changed significantly since the rise of the internet.

If Bigfoot is a large primate, what environmental conditions would it need to survive? Seems like they hold up remarkably well in many different environments, from rainy woods, snow, and desert conditions, as I just interviewed a man who saw one in the desert at night in August of 1987.

How would you explain the elusiveness of Bigfoot? The witnesses have often reported something as fast as a deer, if not faster, so that physical trait might explain a lot when it comes time for the exit strategy when Bigfoots encounter humans.

What are your current research goals and how do you go about them? Research goals, for the time being, are trying to find current reports and to assess the validity of the witness(es) involved.

What do you feel is your strongest skill/talent that helps you in this field? I feel I am a good researcher; I try to turn over every leaf to uncover what the real story is. For instance, there is a story I shared in the *Bigfoot Times*, where there was an individual that was publicly critical of Todd Neiss' sighting in 1993. That person stated that Todd could not have seen all the detail that he claimed to. Although in fact as you start to turn over all the leaves, at no time did Todd say that he saw a tremendous amount of detail. What he saw was not a person and was jet black. This is a case where the eyewitness never stated the information that the researcher claimed and may have been overly opinionated about an account that he doesn't seem to know too much about.

It is important to stick to the facts, when, where, what time of day, etc... Canadian researcher, Tom Steenburg, who is still active, is of that opinion.

If you could get better at one research-related skill, what would it be and why? I would like to have access to more data and time. I have realized that this is not a one-person enterprise, there is too much data out there and to wade through it all as one individual would be impossible in a lifetime. I feel I have stayed consistent with my research techniques, asking questions from eyewitnesses.

What are your current projects? Monthly, the Bigfoot Times, bringing a readership up to date on all things Bigfoot. At this point in my life, I have the largest physical files on Bigfoot than anyone else in the world. I have perhaps access to some information that many other people may not have access to. In that sense, I can't say that I am cutting edge, I just have access to more data than most people. The people who follow this subject closely know who I am so I say that's good enough.

What is your biggest accomplishment in this field? Perhaps the discovery of the authorship of the original newspaper article on the Patterson-Gimlin film, which was printed one day after the film, on October 21, 1967. No by-line accompanied the article. Years later, through microfilm research at the Humboldt County Library, I cracked the case. The author of the piece was Al Tostado.

What is your elevator pitch for open-minded skeptics? It is an open question; on one side you have the "believers" and the people who are the "knowers" (the ones who had a sighting and there is no mistaking what they saw). On the other side, you have the doubters and skeptics who insist that there is nothing out there. I would say, take the middle ground, no one has proven either way that such a species as Bigfoot in North America does not exist. The moment someone says that they don't exist, the next day a credible eyewitness has a sighting. That's how the subject goes and that's how I would tell the skeptics and doubters to approach it and keep an open mind. For example, if you look at the field of astronomy, there are many things that we pre-supposed and as years have gone by, the whole idea of the universe has changed from being a sun-centered universe to an earth-centered universe, to the fact that we are just the pale blue dot purged in the middle of the Milky Way and we are really not that important at all.

What are your must-read books for skeptics and new researchers? Seems like my answer changes as the years go on. Ivan Sanderson's *Abominable Snowmen*, 1961, to get a feel of the global nature of the mystery; John Green, *Sasquatch: The Apes Among Us*, 1978, to get a feel of the North American reports PRIOR to the internet; Marion Place, *On The Track Of Bigfoot*, 1974, for the younger reader, a great introduction to the subject; Rick Berry, *Bigfoot On The East Coast*, 1993, to get a detailed picture up to that time of all the reports on the East Coast, which may have exceeded the West Coast activity at that time.

Why do you think there is not more evidence or proof? That is a good question, I think possibly because the numbers are so low. If you think about it, we see some deer in our environment although not nearly that many considering their population. There may be a million deer for every 30 Bigfoot for instance. Evidence is collected and gets lost over the years, some because people don't think much of it. For example, some of the hair that was collected and lost from the Bossburg incident could have been examined with today's technology and been the "golden hair". There is enough evidence to keep you going as a researcher.

How do you feel mainstream scientists will publicly accept the existence of Bigfoot? In my mind, the only way they are going to accept a creature of this extraordinary disposition, we will need robust physical evidence in the sense that a body is collected. The fact that the subject alone conjures up images of Santa Claus and unicorns, etc., some already think it is a mythical creature. I don't think at this time anything else will suffice on the matter. Both Dr. Grover Krantz and John Green were also of that opinion even back in the 1970s.

What suggestions do you have for researchers who want to get involved in this field or further educate themselves? I would say 2 things; when you go into the field, be certain that you have a camera and a notepad, so you can document what has transpired if you ever find anything. When things are written down, experiences become ingrained in your memory at the time of an occurrence. You never know when that "kodak moment" is going to happen. Roger Patterson in 1967 had the presence of mind to understand that concept, he told René Dahinden that he practiced pulling out his camera from his poncho and the saddle bag on the horse, just in case. Since he did this, he was able to grab the camera before the horse took off on that day. Without the camera, that event would have been another couple minute eyewitness sighting and instead, it became the film of what he saw.

How to follow Daniel Perez:
www.bigfoottimes.net

TODD PRESCOTT

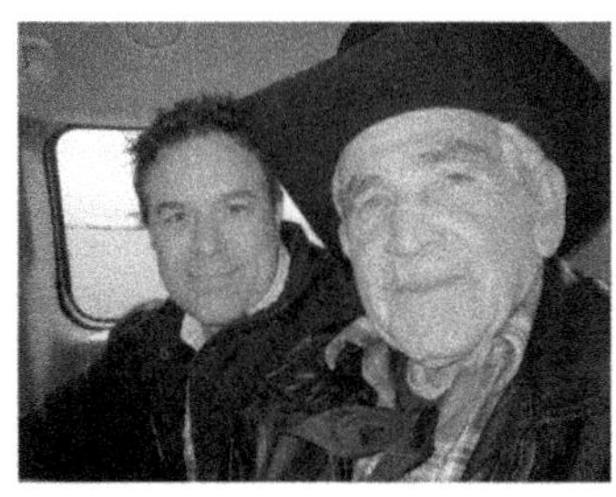

Todd Prescott grew up on the edge of a forest on the Bruce Peninsula in Ontario, Canada. Countless hours exploring that forest and others in the area made him extremely comfortable in wilderness settings. After high school, he moved to Toronto to pursue his interest in music, and he earned a degree in Jazz Studies. In addition, he is also a professional journalist, and he has transitioned that expertise to his research on the subject of Bigfoot.

I reached out to Todd about this book because of the impact he continues to have on the research history of the subject. A few Bigfoot enthusiasts may not be aware of his work. He has been collecting and archiving books, newspaper articles, videos, and interviews from researchers such as John Green, Peter Byrne, Grover Krantz, René Dahinden, and Larry Lund. One of the things I appreciate most about Todd and other historians of the subject is the journalistic approach of uncovering and sharing documented events and not creating new theories or hypotheses.

In addition to enjoying getting to know him, Todd has been supportive and an excellent resource. Along with his professional work as a writer, research experience, and vast archival materials, he is helping to preserve the subject matter and giving generations to come an avenue to explore the history of the Bigfoot phenomenon.

Can you share an embarrassing or funny moment from your childhood? We used to do horrific things like throw snowballs at cars. We would hide behind the Bell Telephone building and whip snowballs at cars for fun. I remember many years after that, I was in my thirties, and I was driving back to the village I grew up in—a village of three hundred fifty people. I'm driving on the road on my way to the village and suddenly I hear something hit my car. I look at the back window and it's the imprint of a snowball. I stopped my car ready to yell at the kid. Then I thought, "Nah, that's karma." It scares the living crap out of you when you're driving, and something smacks your car. So now I know how all those drivers felt. Karma.

What is something that most people do not know about you? Most people do not know that I studied Mixed Martial Arts and used to write for various magazines on the subject (including *Ultimate MMA* and *Elite Fighter*). For years I had monthly columns and wrote dozens and dozens of articles. While involved in that endeavor, I had the honor to interview many well-known UFC fighters as well as train with several of them.

What was an embarrassing or funny memory while researching or any moment in your past? While in the mountains of Bella Coola (B.C.), I was certain that I was about to be attacked by a mother grizzly or black bear. When walking in some tall grass, I heard what I thought was a bear cub remarkably close to me calling in distress for its mother. I readied my huge canister of bear spray and oversized hunting knife while I braced myself for the mother's charge. The distress call continued for what seemed forever while my heart beat double time. As the seconds dragged on, it became clear that a mother bear was not going to attack. I crouched down to see if I could see the baby bear through the grass. Much to my surprise (and relief), there was a mother grouse trying to lead me away from her nest. My life flashed by me because of that damn grouse. Thankfully, I was by myself, and no one knew that I had mistaken a mother grouse for a baby grizzly bear. I suppose now, everyone knows.

If you could take one person past or present on an expedition with you, who would it be and why?

It's tough to pick just one. I did do some field research with John Green back in 2013. Had I not, I would have chosen him. I would probably have to pick Roger Patterson then because there are so many unanswered questions concerning the Patterson-Gimlin film.

When and why did you get involved in research? I think most young boys are fascinated with the unknown. I grew up in the 70s and 80s. Bigfoot, UFOs, and all sorts of mysteries were all over TV shows and magazines. My school library had a great section on Bigfoot, Loch Ness Monster, and other things of that nature. Of course, those topics seemed more interesting than reading novels. It sparked something in me, and it seemed like something that would be an interesting pursuit for me. I could do it in my spare time and it wasn't an expensive endeavor. I've always liked the outdoors, so why not couple that with Sasquatch research? Of course, I was also traveling in bands in my late teens, early 20s, and it just made sense to get involved with the study. I had the entire day to myself, and I was only working three or four hours a night, so why not use my off-time time wisely?

Have you interviewed many witnesses? I've interviewed countless witnesses. With some of them, it was definitely questionable what they experienced—it may have been more of something that they thought they experienced, or they heard something and automatically presumed it was a Sasquatch rather than looking at other options. Some witnesses want so badly to believe that it's a

Sasquatch. They'll say things like, "Well, it was a Sasquatch mimicking coyotes," or, "I know it wasn't a person." So, you absolutely must have a filter when you're dealing with people who are making these claims. You really need to have a "BS" detector. I've talked to people though who have come face to face with these things, and it would seem that they're not making it all up. Some of these people have a lot to lose by going public with their sightings. I remember years ago I interviewed a lady who was a recognized bird expert in Canada—an ornithologist. One morning, she and her husband watched a Sasquatch for a while behind their cabin. I did a follow-up at the site and spent a night there at her cabin. Unfortunately, nothing happened to me, but the witness' credibility and observation skills made it difficult to dispute what she claimed to have witnessed. But following up with witnesses and doing on-site investigations can be like chasing ambulances—the event has already happened. It's gone and you can't expect to have another experience in the same spot.

If they exist, what do you feel Bigfoot/Sasquatch are and why? In my opinion, Sasquatch would have to be an as-yet classified hominid. Their reported behaviors and morphology strongly suggest that they are in the great ape family just like humans. So many people nowadays are saying that Sasquatch are not apes but are of the human lineage. This of course is a misnomer because humans are apes—great apes to be exact. Therefore, if someone is of the mind that Sasquatch are humans then Sasquatch absolutely are apes.

What are your current projects and how can people follow your work? For the past five years or so, I've spent a considerable amount of time uploading videos to my YouTube channel, *The Sasquatch Archives*. Aside from that, I continue to chip away at some books I'm working on. I'm also currently editing a mammoth book about the early period of Sasquatch research.

What type of historical information do you have? I have old scans of articles from newspapers dating back to the 1850s. As far as original stuff, the oldest thing I have in my collection is a first-edition book from 1888. There is a section in there that talks about something that basically invaded a camp—something on two legs, something that they didn't have a name for, but based on the description, could have been a Bigfoot or Sasquatch. I've got a few other first-edition books from the late 1800s such as *The Hermit of Siskiyou*. I've got old magazine articles dating back to the early 1900s that talk about things that could be Bigfoot or Sasquatch. Of course, the name Bigfoot and Sasquatch wasn't around in the early 1900s, so often they were called Wild Man, Hairy Man, or other local names in the earlier publications.

How did it start? I began collecting more and more books and magazines pertaining to Sasquatch research, and soon I had hundreds in my collection. I had been making connections with other researchers around the world through the years. I eventually began contact with long-time researcher John Green. John invited me out to visit him in Agassiz, B.C. and permitted me to go through his extensive files including hundreds of letters, photos, slides, negatives, and other odds-and-ends on the study of Sasquatch. I spent 3 years visiting with John and archiving much of his collection. In 2015, I started my YouTube channel, *The Sasquatch Archives*, to showcase unique and rare content about the study of Sasquatch, including materials from the Green collection. The channel will continue to grow as other research treasures are uncovered and eventually uploaded to the channel. Working with researcher Christopher Murphy, I helped to edit some of his published materials and his online newsletter. I worked with Dr. John Bindernagel on some of his research prior to his passing, and I co-edited one of Dmitri Bayanov's books (*RUSSIAN HOMINOLOGY The Bayanov Papers*). I've written Sasquatch articles for magazines such as *Fortean Times*, *Atlantis Rising*, and *The Backwoodsman*. As of now, I am currently editing a book on the early years of Sasquatch research that is due out in 2022. I too am working on books about the subject matter which will hopefully see the light of day at some point.

Do you feel since the 1960s that we have gotten any closer to proving the existence of Bigfoot and why? At the present time, no, I don't. For scientists, the only concrete evidence that will prove the existence of Sasquatch is a specimen. Everything else to them is moot. All we have since the 1960s is a lot of hearsay—people claiming this and that. We may have more footprint finds and more reported sightings, but to prove the existence of Sasquatch, we need more than just that. We can say what we want about Sasquatch, but in reality, the only Sasquatch expert is a Sasquatch. Those claiming to know this and that about Sasquatch are only fooling themselves.

How to follow Todd Prescott:
YouTube: The Sasquatch Archives

MATT PRUITT

You will not find a nicer and more genuine person than Matt Pruitt. He has been researching the sasquatch phenomenon since 2002, and subsequently began conducting field research in 2004. His pursuit of this phenomenon has led him across the continent; primarily focusing on regions in the southern Appalachians, the Pacific Northwest, and the US Interior Highlands. Matt spent seven years as an investigator for the Bigfoot Field Researchers Organization (BFRO) and is currently on the Board of Directors for the North American Wood Ape Conservancy (NAWAC). In addition, he is the producer and editor for the podcast "Bigfoot and Beyond with Cliff and Bobo".

Matt Pruitt Introduction

When I created the concept for this book and shared the idea with Dana, Matt was the first person I called to get an opinion. Not only has he become a trusted friend, but he also has an innovative and intelligent mindset. His professional background has equipped him with experience that plays a vital role in his research and collaboration with his peers.

Though he has not had a visual observation of a sasquatch, he has had experiences over the years that were consistent with the phenomenon as described by the testimonial claims of sasquatch witnesses. In addition to his field efforts, he engages in numerous public-facing discussions about the subject in various media and at speaking events. Matt has also interviewed thousands of witnesses. I encourage anyone who is interested in the subject of Bigfoot to take the time to listen to Matt's theories and perspectives on the phenomenon.

Where did you grow up? I grew up in Northeast Georgia, first living outside of a little town called Blairsville. When I was five, we moved near the small town of Helen, where I spent most of my youth in the foothills of the southern Appalachian Mountains.

What are your hobbies? I've always been interested in the outdoors, but my other passion is writing, playing, and recording music. I started playing guitar and drums by learning to play my favorite songs and later began writing my own songs in my early teens. I began playing music live in clubs and bars when I was 15 years old; that was 25 years ago this year. I later ended up in a major label recording band called Injected out of Atlanta, Georgia. I continue to write, play, and record music here and there.

Who are your favorite musicians or what type of music are you into? My favorite band is The Beatles and a lot of the artists that I love stemmed from their style of rock or pop-rock music. Queen is one of my favorite bands, and Freddie Mercury is probably my favorite vocalist of all time. I was a teenager in the nineties so that music had a big impact on me. There are a lot of nineties bands that I absolutely love; the Smashing Pumpkins, Stone Temple Pilots, Nirvana, Weezer, and Soundgarden, to name a few. In fact, I have a nineties alternative rock cover band here in Nashville. I also adore the music of John Mayer, Butch Walker, Jeff Buckley, Fiona Apple, Ben Folds, Chris Thile, Dawes, Jellyfish, and many more.

What is a bucket list item that you would like to accomplish in the next 6 months? I certainly would like to see a sasquatch within the next six months!

Why did you get involved in the subject of Bigfoot? I had an experience in the summer of 1999 that was consistent with many claims associated with the sasquatch phenomenon. It was primarily an auditory experience; we never saw the things that we heard because it was at night. It was extremely dark, as it occurred at night in the middle of the summer in dense vegetation. We only had small flashlights, so we couldn't see particularly well. At the time of the event, we didn't know what was occurring, just that it was very frightening. We were certain that it wasn't other people or a known animal that we were encountering.

A couple of years later, I found the Bigfoot Field Researchers Organization (BFRO) website among some other resources. As I was reading reports, I genuinely remember laughing because the thought of it was so preposterous. As I read more reports, however, I found that certain testimonies described what we had experienced to a "T".

At first, the question as to whether or not sasquatches existed was one that I thought I could answer for myself in no short order. I assumed that I would come to a personal conclusion, and that would be that. Now, over 20 years later, I still am seeking that conclusion.

Why is it important to understand the history of the subject? Without an intimate familiarity with the current state of knowledge regarding a given subject, one can't even begin to know which questions should be asked. Understanding the history of the sasquatch phenomenon, the claims and evidence associated with it, and studying the history of the pursuit is crucial to understanding what questions need to be asked in order to arrive at a solution. People neglect to take that step at their own peril. They risk missing necessary information, often leading them to premature conclusions that have already been soundly dispelled.

In addition to understanding the history, what do you feel is most important for new researchers? Defining your expectations is paramount. Knowing exactly what it is that you want to accomplish will dictate the nature of your pursuit. Once you've properly defined what it is that you aim to achieve, you must conduct yourself accordingly in order to accomplish the goals that you've outlined.

What was an embarrassing or funny memory while researching or any moment in your past? During an impromptu field outing in the eastern Cascades of Washington, myself and a trusted field partner both missed seeing two upright creatures that were witnessed clearly by four of our companions; all of whom were previously unconvinced that sasquatches could exist. I've never forgiven myself for missing that opportunity!

If you could take one person past or present on an expedition with you, who would it be and why? It'd have to be my father, who has accompanied me on field surveys in the past. I'd like to facilitate an observation or experience for him, and it would be especially meaningful to experience that together. It goes without saying that we inherit certain traits from our forebears, and I've undoubtedly gotten much of my curiosity about the natural world from him.

If they exist, what do you feel Bigfoot are and why? In my estimation, the most parsimonious hypothesis for the existence of the sasquatch is that members of the Asian ape lineage that produced the known genera *Indopithecus* and *Gigantopithecus* extended their population into North America.

If Bigfoot is a large primate, what environmental conditions would it need to survive? Dense forests, an abundance of food and water, and access to other individuals in nearby habitats for reproduction.

How would you explain the elusiveness of Bigfoot? Large, rare animals are very difficult to encounter and observe. They require large home ranges and are often highly mobile within those ranges. There are a host of other mammals that are similarly elusive. Sasquatches don't seem to be completely undetectable, given the number of observational testimonies offered by credible observers.

How did you get involved with the North American Wood Ape Conservancy (NAWAC) and what are their goals? I applied for membership in 2016, having met a couple of their members previously in 2010. I was always interested in their work and felt like I could contribute in some way.

The NAWAC's primary goal is to validate the hypothesis that a biological ape species is at the root of the sasquatch phenomenon. Over the last few years, our efforts have expanded a great deal; especially with attempts to obtain video and photographic evidence. Visual data might not constitute the ultimate form of "proof", but in terms of providing supportive evidence for the ape hypothesis, we do believe that visual data would be impactful. The pursuit of visual data comprises a large part of our efforts currently.

How long were you with the BFRO and what impact has it had on your research? I was a member of the BFRO from 2007 until 2014, and my involvement with that organization had a huge impact on me.

As an independent researcher in the early 2000s, it was extremely difficult to find other people to communicate with who were pursuing the sasquatch phenomenon. When I joined the BFRO, I suddenly had access to a large network of people who were all engaged in the same endeavor. Given my youth and experience level at the time, my membership allowed me to learn from many people who had been conducting field research for a lot longer than I had been.

Since the BFRO had the largest presence of any source for sasquatch information online, they received a tremendous number of eyewitness reports. That gave me access to literally thousands of witnesses. I've interviewed over 2,000 witnesses to date, and a great many of those came to me through the BFRO. I also made a lot of lifelong friends (and even met my wife) as a result of the organization's efforts. I had many positive experiences during the seven years that I was a member, and I wouldn't trade those for anything.

In one or two sentences, how would you describe Cliff Barackman? Cliff is the hardest working person in the history of Sasquatchery. He's a great friend and is such a great exemplar for how to represent the subject in a way that's informative, respectful, and engaging.

How about James "Bobo' Fay? Bobo is the most authentically unique individual I have ever met. He is purely himself in a way that is inspiring. There is no persona with Bobo; what you see is what you get. That authenticity is precisely why so many people like him.

What impact do you think *Finding Bigfoot* on the subject? It had a huge impact at multiple levels. I think it brought about the ability to have a conversation about the subject in public discourse because it brought it further into the context of pop culture. When a show like *Finding Bigfoot* is *that* successful, it filters its way into a lot of different conversations and dialogues. I think it made a lot of witnesses feel more comfortable about sharing their experiences. It also generated a certain number of fabricated reports by attention seekers. Nevertheless, I think it's had an overall positive impact.

Matt Pruitt on Finding Bigfoot

How do you feel mainstream science will publicly accept the existence of Bigfoot? Most likely as the result of a specimen (or a significant portion of one) that's collected. That might be facilitated by an individual dispatching a living animal, or by the recovery of one that's died in the last several years or decades, or by the recovery of a bone from a sasquatch that expired at some point in the modern era. I also suspect that acceptance will happen in minimal necessary increments. For example, if someone recovers a sasquatch specimen on the Olympic Peninsula of Washington, I wouldn't expect scientific, academic, and governmental institutions to immediately accept that sasquatches occur in many other regions of North America. Acceptance and recognition will occur locally around the region where the initial discovery is made and will undergo a slow process when it comes to identifying other populations elsewhere.

What's your elevator pitch for open-minded skeptics on the existence of Bigfoot? We still have no consensus about what generates the claims and evidence that comprise the sasquatch phenomenon. It's easy to dismissively state that it's all the product of the human mind, which is what it would have to be if there's no such animal being observed and encountered. The more that you familiarize yourself with the history of the phenomenon and the associated evidence, the less likely that it seems that all of it could be explained away by some function of the human brain operating without actual empirical referents in the environment. To paraphrase John Green: Either these animals are real, or they're not. There's no in-between. If they are real, then *some* of these direct observation claims are probably true, and *some* of these tracks were probably left by the animals described by the claimants. If no such animal exists, then *every* direct observation claim has to be fabricated, along with *every* track. Occam's Razor cuts on the side of the sasquatch being a living animal.

How to follow Matt Pruitt:
mattpruittonline.com
Twitter: @mpruittbigfoot

PHOTOS AND INFORMATION II

Squatchermetrics analyzes documented historical reports and provides commonalities and differences based on the data points.

North American Bigfoot Reports

1. **Total number of reports**– 7,616
- Date range - 1811-2022
- Most recent report – Jan 2022, Humboldt County, CA

2. **Visual reports** - 4,137
- Date range - 1869-2021
- Most recent visual reports – December 2021, Lewis County WA and Surry County VA

3. **Vocalization reports** - 2,026
- Most common states (in order) are WA, CA, TX, OH, and PA.

4. **Object throwing** - 380
- Most common states (in order) WA, KY, MI, VA, and OH

5. **Multi-person** - 3,844

6. **Reports where children were present** – 172 (children playing, which is the closest I can get)

7. **Road crossings** – 1,160
- Most common states – WA, TX, FL, PA, and OH leading the way.

8. Who reported - hunters, hikers, homeowners, law enforcement, forestry, bigfoot researchers, etc.

• In order it's Driving, At Home, Hiking, Hunting, Camping, Other and Fishing.

9. Near homes - 1,300

• WA, OH, PA, TX, MI, and VA in that order.

10. By season in order - Summer > Fall > Spring > Winter

11. Reports by month in order - August > October > July > September > June > November > May > April > January > December > February

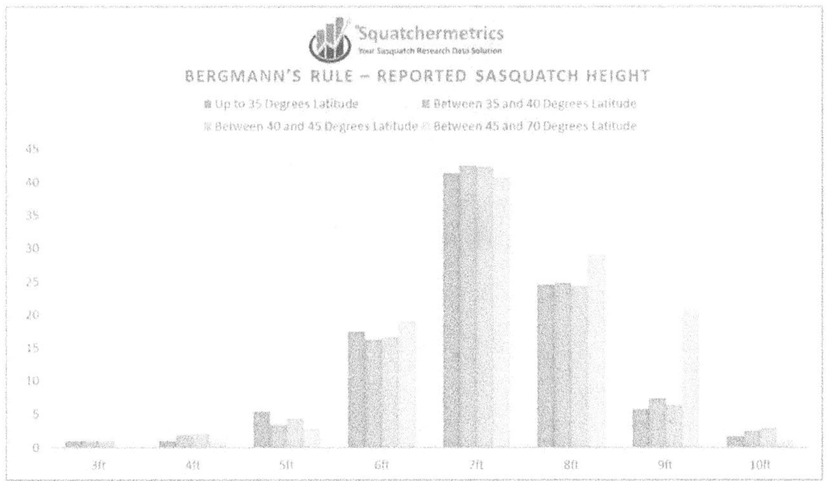

The previous chart demonstrates the reported heights of Sasquatches based on the *Bergman's Rule* theory that species further away from the equator and in colder climates are larger to adapt to their environment. Notice the increased number of 8ft and 9ft visual Sasquatch reports between 45 and 70 degrees latitude.

Here shows clusters of reports from 2017 through 2021. Photos courtesy of Squatchermetrics.

Squatchermetrics Images and Data

Casting Suggestions from Tom Shay

Casting is an important step in collecting evidence. There are some who feel that their casting skills are not great and decide not to cast anything while researching. At one time I felt the same way. I remember meeting Charlie Raymond and showing him my casts for the first time. Some were broken and I used clear packing tape to hold them together. Others looked like oatmeal; they were just not done well. Over time I practiced and honed my skills. Those first casts are still in my collection with the tape still attached. I continue to proudly display them. We all have our own technic and skill set although some consistencies are shared by all individuals who cast tracks on a regular basis.

Before casting a track, document the print with photos. Capture every angle, this will typically consist of 15 to 20 photos. Take depth measurements of the toes, and from the ball of the foot to the heal. After retrieving the casting, you may want to wrap it in several layers of newspaper. Change the paper every 4 days, this helps cure the cast evenly. Casting should be fun just take your time. The more you cast the better you will get.

There are environmental variables to consider when casting, such as the saturation of substrate humidity. In addition, there are a variety of suggestions on the type of casting material to use, the mixing ratios, and techniques. Research methods from individuals and organizations that cast prints, consider your particular environment, and practice!

This image is from the report that Diane Stocking investigated in Titusville, Florida. Two large oily handprints were left on the glass window at an interesting height. Photo courtesy of Diane Stocking.

The author and Loren Coleman at the International Cryptozoology Museum in Portland, Maine. Photo courtesy of the author.

Tom Shay's Nutella trap. This was hung from a tree and the jar was placed inside. The openings of the pipe were lined with sandpaper to capture hair. Photo courtesy of Tom Shay.

Panel Discussion at the 2022 Ohio Bigfoot Conference. Left to Right: Matt Moneymaker, Adam Davies, Dr. Jeff Meldrum, Cliff Barackman, Ranae Holland, and Charlie Raymond. Photo courtesy of the author.

Daniel Perez and René Dahinden, August 31, 1980, in Richmond, British Columbia at his home at the Vancouver Gun Club. Courtesy and copyright © Daniel Perez, 2022.

Dr. Grover Krantz in his garage with his collection of skulls and molds of Sasquatch tracks. Photo courtesy of Rick Noll.

John Green in front of a few pictures of Sasquatch tracks he had collected. Photo courtesy of Rick Noll.

DR. ESTEBAN SARMIENTO

I am thrilled to include Dr. Esteban Sarmiento in this book. He is unique compared to many of the others I interviewed based on his skepticism that Bigfoot exists as a physical and biological entity. He is a primatologist and biologist and noted for his work in great apes and African primates. His primary field of study is hominoid skeletal biology of both extinct and living species. He is a contributor to the *Gorilla Biology textbook* which is currently part of the Cambridge Studies in Biological Anthropology program. His work has been cited in the Harvard University program and in a multitude of other University curricula.

He is fascinated by the Bigfoot Phenomena. Over the years he has been called in as a great ape expert by various Bigfoot researchers. He is featured in *MonsterQuest* and *Sasquatch: Legend meets Science*. I enjoyed discussing Bigfoot with Esteban. He is genuine, candid, and funny. I was impressed that he could be skeptical but not closed-minded to the mystery. He shared an essay with me that he authored on the subject which takes an objective view on the topic and takes an interesting position on the phenomenon that should be considered.

For the past two decades, Esteban has researched several historic events related to Bigfoot alongside individuals such as Doug Hajiceck, Dr. Jeff Meldrum, and Richard Noll. His objective stance as a scientist and skeptic has been essential to maintaining a credible balance to theories on the possible existence of Bigfoot. He has played an impactful role in the history of this phenomenon.

What are your hobbies? I sail, bicycle and coach wrestling, so I keep my body moving.

What's your favorite type of music or musicians? I'm pretty eclectic, I'd say the 60s and 70s, The Hollies, Creedence Clearwater Revival and Johnny Cash. I'm a big Beethoven and Rimsky-Korsakov fan and enjoy jazz. One of my neighbors is Dizzy Gillespie's daughter, a jazz singer in her own right, albeit now retired.

What was a funny childhood memory? I remember doing a lot of funny things, when I was young, I was a jokester. I must have been 11 years old and I climbed the Lincoln Memorial. It's actually quite tall, I got on top of his lap. It was difficult, I had to grab the marble pleats in his pants. the guard was trying to talk me down and I said, "I don't understand English" [laughing].

What was an embarrassing moment in your past? Stepping knee-high in horsesh*t when I was two years old.

What is a bucket list item that you would like to accomplish in the next 6 months? Finish authoring my book on human evolution and the fossil record.

What type of research are you currently involved in? My main research is not related to Bigfoot. My work is focused on African primates, As regards the latter I want to return to Cabinda Angola where the most southerly distributed gorillas live to continue my work with them.

If you could take one person past or present on an expedition with you, who would it be and why? Steve Fox, my paleontology and stratigraphy professor at Rutgers College. He knew all the best priced bars and food joints in North America and would never stop to eat at a fast-food chain i.e., McDonald's Wendy's, etc., which he despised.

If they exist, what do you feel Bigfoot/Sasquatch are and why? I don't know what Bigfoot are. I don't know if it is a real living mammal closely related to humans and great apes or a figment of the human imagination. More physical evidence could serve to clarify what it is.

How would you explain the elusiveness of Bigfoot? Familiarity. Humans are clearly not familiar with the signs that they leave behind. An alternative explanation is that Bigfoot is not a physical entity.

How did you get involved with Bigfoot researchers? My interest has always been great apes, I completed my PhD thesis on orangutans. Sadly, once you understand their environment, you realize the land that they're living on is going to disappear in a very short time. I decided to move to Africa in the late 80s and study African apes and their fossils. I would come back to the American Museum of Natural History every other year. Doug Hajicek once came to the museum asking around about great ape researchers that were interested in Bigfoot. Somehow, he bumped into me. He asked me questions about great apes and mentioned that he wanted to do some type of movie. That's how it all started. Once he made the movie and it was released on TV, more people started getting in touch with me. I credited Doug Hajicek with being the one individual that did the most to bring Bigfoot into the mainstream consciousness of the American public.

Was there any potential evidence that you found compelling? With MonsterQuest and also the film, we went to many interesting places. I remember visiting a fishing island in Western Ontario. There was a Bigfoot sighting, not very far away. We went to investigate and that was interesting. To talk to the witnesses, they were First Nations people. Here in the USA, we would call them Native Americans. They swore that they saw a Bigfoot and showed me where it walked. The trail was only two days old. It was a common animal trail, nothing specific to Bigfoot. I remember I put my glasses on and I looked through all the bushes and plants to see if there was any hair. I did find some, although most of it was moose and elk, we also found hair/fur that seemed to be from wolves and bears, among others.

I also went to northern Washington, to the Cascade Mountains with Doug, to film and investigate the Skookum Cast. That was also a very interesting place. One of the things that I have noticed, is that where Bigfoots are usually sighted are also areas of high animal concentration. That means there is an abundance of other animals in the area, in fact, the Skookum Cast, if you ever looked at it, has many tracks of other animals that are identifiable. It's clear there are traces of coyote and mule deer, and I even found what looked to be like mountain goat hairs. From the cast, I couldn't tell if it was a Bigfoot sitting there or not.

Basically, as far as the Skookum cast, I put on my glasses and picked hairs. If you have good glasses, you can see some of the differences in hairs without having to bring samples back to a lab. I was able to visually distinguish hair and sort out some of the diverse types of animals. I then examined several of the hairs through an electron microscope, and that's always the most exciting part. I even had a couple of my own hairs in there because they fell off my head when I was looking. Of course, I knew right away who they belonged to. [laughing]

Is there any piece of evidence that is compelling enough to call for scientific research? The question of Bigfoot is a complicated one. All of science deals with reality, whether it's physics or chemistry, every science deals with it. This is a result of the shortcomings of our sensory organs which no matter how good they are actually filter reality and do not precisely report it. For the people that see Bigfoot, it is no doubt real to them. To me, it is not yet real because I've never seen it or experienced it. So, what I think is the most compelling bit of evidence are all the eyewitnesses, and not just in our country, but in places all around the world. That means something. We just don't know what it means. We are dealing with either a real physical entity or some kind of reality issue.

Dr. Esteban Sarmiento Audio

Esteban Sarmiento and Richard Wrangham. Photo courtesy of Esteban Sarmiento.

Are there any similarities between the Bigfoot witness accounts and the African apes that you have researched? Many witness testimonies jive very well with African and Asian ape behavior, Great ape behavior overall. Wood knocking and throwing objects are things that chimps do. All Great apes do throw objects, some more than others. Orangutans will throw, gorillas will as well, but less often.

What do you think about the First Nations and the Native American accounts of possible hairy creatures? Like I said, the most compelling evidence for Bigfoot are all these eyewitnesses. It's not just eyewitnesses presently in the here and now. There are eyewitnesses going back three or four hundred years. Not just in North America but also in Asia. It's also part of the culture of many human groups, so in that sense, it's really hard to divide the two (reality and perception). Because, in some ways, if it is part of their culture these humans expect to see it. Since it is part of their culture, it makes one think that there must be some aspect of reality to them. To me, currently, since I have not experienced Bigfoot, I am fixated on the reality perspective. To those that have seen it, or believe they have seen it, it's real, and reality is not an issue to them. But this isn't necessarily so.

What do you think about the Patterson Gimlin Film? I love the Patterson Gimlin film because if it was faked, the person really knew how far he could take it. It doesn't show enough resolution for anyone to conclusively say it's a fake. At the same time, it doesn't show enough resolution for anybody to be certain that it is an otherwise undiscovered animal. That's my opinion, other people can say, "He's full of sh*t, that's obviously Bigfoot". I measured Patty from the film and authored a paper on Bigfoot. The basis for the paper can be summarized as follows, Bigfoot has some reality in its biogeographic distribution, which means if there was/is a great ape in the Americas, it must have come from Asia through the Bering Strait. So as far as that's concerned, what are the most likely bigfoot sightings given their geographic locations? Could Bigfoot be in Hawaii? Could it be in Australia? Is that a biogeographic possibility? The paper discusses that.

If it isn't, and some of those sightings (i.e. Hawaii and Australia) are something imagined, what could it be? I tried to propose an explanation of what it could be if it's not something real. To me, it is not easy and much less fair-minded to take the thousands of eyewitnesses' accounts and say, "Oh, these people are just lunatics". To others, it may be very easy to say that, unless of course, you are one of the lunatics. Then and there it stops being easy. You can't just dismiss all these people as crazy. There must be something there. It's fascinating, At the moment I'm more interested in why people see it than in anything else.

What do you think about all the footprints that are found? On the one hand, if they have such great footprints, why can't they get another part of them? If I track gorillas, say which I have for a long time, I can always find hair, feces and occasionally even a fetus. When I have looked at Bigfoot prints, I didn't see any of this. Even within the footprint, I should be able to get hair. Many times in gorilla footprints, I can find hair.

Similarly, I question the validity of the possible Bigfoot nests that have been discovered. For example, If I find a gorilla bedding area, I can see they sleep in them and defecate in them, and I can find hair in them. The Great apes all make nests each night, and occasionally during the day. We can find examples of vegetation that an individual chewed that night while in the nest. I remember once finding in a chimpanzee nest, a porcupine that had been killed and eaten. You could tell by the way the limbs were broken. The chimp turned them around and around to try to pull them off and had eaten all the meat. These are things that you find in primate environments, but it's yet to be found for Bigfoot.

What will it take to prove the existence of Bigfoot? If it exists, it will take a body. We live in the age of jive, where everybody is lying because they could make a buck from it. I'm sure even if we find a body, some people, still wouldn't believe it. The question of Bigfoot existence is that polarizing.

THOMAS SHAY

Thomas Shay is the founder of the Northern Kentucky Bigfoot Research Group, which he started in 2013. His quest to find proof that these elusive creatures exist was fueled by his first sighting he experienced in 1987 while he was home on military leave. Many consider him one of the top researchers and trackers out in the field today and Thomas currently owns the largest collection of original Bigfoot casts east of the Mississippi.

One of the most unique pieces of evidence that Thomas has collected is the *Nutella* jar finger impression casting. He and his research team built a trap using a large PVC plumbing tee fitting lined with sandpaper to catch hair samples of any animal that would reach in. They then placed an unopened jar of Nutella inside of it and strung it between trees. When they returned approximately a month later to check the trap, it was disturbed and the jar was 30 feet away uphill. Inside the jar, it was clear that something had reached in apparently to scoop some of the Nutella out. Thomas placed the jar in a portable cooler to transport, and when he returned home, he placed it in his freezer. The next morning, it was frozen solid which allowed him to cast it. The result: four visible impressions and two of them showing what appear to be fingernails. This was certainly a fascinating discovery by Thomas.

Nutella Cast Images

Along with his extensive collection of casts, Thomas has collected a massive amount of added possible evidence through his diligent fieldwork. Getting to know him, he is passionate about his research and very humble. Thomas does not bring attention to himself, instead, he focuses on the bigger picture, the preservation and understanding of the Bigfoot species. He has educated many enthusiasts and fellow researchers over the years and continues to play a significant role in the Bigfoot community.

Which current or past scientists/researchers have influenced your work? Roger Patterson, John Green, Paul Freeman, Dr. Jeff Meldrum, Cliff Barackman.

What was the most compelling experience or evidence you have encountered firsthand? In 2014, I was working in one of my research areas and I came across footprints. This was mid-day, I got off my 4-wheeler to investigate and noticed that these prints were large. I measured and photographed them. After I poured two castings, I smelled something odd which made me look up. Approximately 30 feet from me, on the other side of my 4-wheeler, was a massive creature. It was huge, the shoulders were approximately 4 feet in width, it was twice of me. It was looking at me, covered in reddish-brown hair with dark eyes. I always pack a handgun, which I left on the 4 -wheeler. We stared at each other, I was on my knees and considered making a run to get my gun although decided against that idea. The creature then looked up, sniffed the air, grunted, and walked on down the ridge.

Next is a funny story, I was so excited; I took out my flip phone, followed it and started filming. I stayed behind it at a distance and followed the creature as far as I could until it was out of sight. I raced back home and told my wife what had happened and said, "I got it on film!". After I downloaded what I recorded, I realized that instead of filming the Bigfoot, I had the camera lens backward and I filmed myself the entire time. I was in tears at that moment.

I went back later with a research partner. Based on our reference points, we estimated it to be at least 9 feet tall and we got some really good foot castings.

If they exist, what do you feel Bigfoot are? A derived branch of man.

How would you explain the elusiveness of Bigfoot? I believe that they have been persecuted and hunted by early man until present day and had to learn to be elusive.

What type of research are you currently involved with? Habituation – We try to figure out how we can make contact with the Bigfoots. If we can study their behavior, maybe we can get close enough to them and somehow communicate with them. Do they have a family structure? Are they clan-like? What type of social skills do they have?

What can we learn most about foot and handprints? We can learn how many individuals are in a specific area; we can decipher the gender and age based on the size and development of the prints we find. Also, based on the bottom of the footprint, and the thickness of the pad, we can understand how developed the foot and individual may be.

What was your first reaction when you saw the Nutella jar removed from the hair trap? At first, I thought we would not find the jar. In the past, we had used peanut butter jars and they would go missing for a couple of days and then show back up at the same spot empty. This one though was found 30-40 feet uphill from where the trap was. This puzzled us as jars don't roll uphill. We did not know at the time what may have removed it from the trap. I picked up the jar, looked inside and noticed something had reached in and pulled out some of the contents, I could see finger patterns and the lid was screwed back on which was interesting. The original placement of the trap was suspended by wire between trees approximately 8 feet in the air. We designed it so no known local animal could get to it unless it was extremely tall. When we arrived, the trap was no longer suspended between the limbs and was tangled up in the branches of one of the trees we suspended it from.

Why do you think there is not more of the type of evidence to prove the existence of Bigfoot? I feel that there is evidence out there that the public has not seen. Economics may play a part in it, if the government acknowledges their existence, land and habitat will have to be preserved and set aside for them. The lumber industry, land development, real estate, and other businesses will be affected.

Do you feel we have gotten any closer to proving the existence of Bigfoot since the 1960s? Yes, with the evidence; footprint casting and sightings.

What suggestions do you have for researchers who want to get involved in this field or further educate themselves? Get out in the woods, be honest, and do the best you can.

How to keep up with Thomas Shay:
http://bigfootlore.blogspot.com/
Facebook: Northern Kentucky Bigfoot Research Group

THOMAS STEENBURG

Canadian native Thomas Steenburg is one of the elder statesmen of Bigfoot. Well respected by enthusiasts and peers, he started actively researching in 1978. Typically referring to the creature as Sasquatch, his knowledge of the history of the subject is staggering and there are few around today like him. He has archived numerous reports, events and methodically studied historical accounts and provided noteworthy insight into them. From early on, he worked closely with and established long-standing relationships with the pioneers of Bigfoot research. They impacted his research techniques and he is one of the few around today that can share firsthand many of the experiences and escapades of the *Four Horsemen of Bigfoot* and other early researchers. Many of these accounts include a touch of comedy and adversity although Thomas found a way to work with all of them without getting thrown in the middle of their controversies.

Many will say that he does the best impersonation of René Dahinden although René hated it.

Thomas has authored 3 books on the subject, *Sasquatch in Alberta*, *Sasquatch: Bigfoot: The Continuing Mystery*, and *In Search of Giants*. He has co-authored 2 books, *Meet the Sasquatch*, with Chris Murphy and the late John Green, and *Sasquatch in British Columbia*, with Chris Murphy.

Over the years Thomas has not wavered on his stance of sticking to the facts, collecting data, and getting out in the woods to perform fieldwork, which he continues today. His investigations are thorough and he executes due diligence to rule out hoaxing and misidentifications. Although he does feel that Sasquatch exist, and may have witnessed one from a distance, he is open-minded to the possibility that they don't exist until proof is found. Whether he is on tv, podcast, or at a speaking event, typically Thomas does not talk much about himself, he focuses on the witness reports, data, and history of the subject. For example, in the 40 years of his research, Thomas will tell you that he may have only come across a handful of trackways that weren't faked and can't be explained, supporting that does not create or exaggerate what he discovers.

For the book, he was generous with his time, overwhelmingly informative on the history of the subject and delivered humor along the way. Thomas Steenburg is a significant figure related to the subject and has influenced the history of Bigfoot.

Where did you grow up? Bancroft, Ontario

What is something that most people don't know about you? Served 7 years, Queen and Country, with the 1st battalion P.P.C.L.I.

When and why did you get involved in Bigfoot research? It started when I was a kid in the mid-1960s, I remember my parents bringing home a large hard covered *Readers Digest* book for education purposes. In this book, there were chapters on tornados, earthquakes, etc., and a large section with colored pictures dedicated to dinosaurs.

In the middle of this section was a two-page article with fuzzy images, I believe it was titled *The Thing in Loch Ness.* I was fascinated and must have read it 50 times.

I then started pestering my parents for more information. We didn't have computers back then so they gave me a library card. I took up everything I could on the subject and while doing so, I came across information about this thing in Canada called the Sasquatch, in the United States they called it Bigfoot and something similar was seen in the Himalayas. Not too long after that, on a school night, I walked into the living room while my parents were watching a movie on the black and white tv. Typically, my father would not allow me to watch on a school night although he knew I was interested in the subject of what they were watching so allowed me to do so. What was playing was *The Abominable Snowman of the Himalayas.* Since I lived in Canada and that's where the Sasquatch was, it made sense to me to make it all about that.

How did you get introduced to the early researchers? The first one I met was Professor Vladimir Markotic in Calgary. He contacted me because he saw an advertisement, I had taken out in the newspaper asking witnesses to contact me if they have seen a Sasquatch. We met and he took me under his wing. We became a team; he was getting up there in age so he performed the academic research and I did all the fieldwork. Through Vladimir, I met Dr. Grover Krantz and others. Later, I was making a trip to the West Coast and decided to phone John Green. He invited me to visit him and that started our friendship. He was an enormous influence on me, he was so methodical with his research.

Around 1980 I met René Dahinden, he happened to be traveling near me and was staying in Lake Louise. He knew that I was doing research and asked me to come see him. He had this green camper truck and I remember approaching it and hearing dishes or something clanging, I knocked on the door and said, "Mr. Dahinden,

Thomas Steenburg here", René responded, "who cares", and that was the beginning of our long friendship.

I did not meet Bob Titmus until 1989, we communicated frequently prior, just never met. For the last 15 years of his life, he battled some health issues so did not get out much. We finally met at the International Society of Cryptozoology Conference when he traveled down with John Green.

I did a lot of work with all the early researchers and somehow never got sucked into their personal wars.

Which researcher was most unique? René without a doubt. René was René there was no way to describe him. He was always feuding with somebody, Green, Titmus – they hated each other. Every time I would see René he would ask, "What does that Titmus have to say about that". I would say, "I am not telling you René, I won't tell him what you said and I won't tell you what he said". He didn't like it although he accepted it and he never held it against me, like he did with so many others.

If you knew the story of how René and John's partnership fell apart, you would be amazed.

Thomas Steenburg Audio

What was an embarrassing or funny memory while researching? Rene Dahinden chasing a bird down a trail. It was attempting to fly off with his bun although the bun was half the size of the bird and could get up in the air with it. Rene finally did catch his bun.

How was the term Sasquatch created? On the west coast of Canada and the United States there were multiple names for them, both First Nation and Caucasian. Before 1929 on Vancouver Island, you would see the term *Mowgli* used often. In the late 1920s, a man by the name of J.W. Burns took a job with the Sts'ailes First Nation people, he was an agent for the *Crown.* He was very sympathetic and the people trusted him, as a matter of fact, he filled in his time teaching at their local school even though he was never a qualified teacher. You can still find some senior citizens today that will say if it wasn't for J.W. Burns, they would of never went to school. The people shared their mythology and folklore and told him about these things in the woods they called *Sas'qets.* J.W. Burns also authored articles that would appear in Canadian magazines, he wrote one that appeared in *Maclean's* on April 1st, 1929. It was titled *Introducing B.C.'s Hairy Giants* and in that article, he misspelled *Sas'qets* and called them "Sasquatch". It's been known as Sasquatch in Canada ever since.

J.W. Burns 1929 Article

How do you ensure a witness is credible? I don't know, if the Sasquatch does not exist, none of them are credible. When interviewing witnesses, there are only 3 possibilities; they saw a Sasquatch, they mistook something for a Sasquatch, or they are lying.

If they exist, what do you feel Bigfoot are and why? A higher primate, it just makes the most sense, assuming of course they actually do exist. I am strictly zoological; I am not into the paranormal and all the "Woo" stuff. In my opinion, that is just clowns chasing their tails. We must prove it exists and until we do so, who gives a damn? We must establish question one before we worry about all the other aspects of their being.

What was the most compelling witness account that you heard first-hand? I have several favorites; stories that are interesting, if they are 100% real, that is the question. The Crandell Campground incident comes to mind. It took place during the Victoria Day long holiday weekend in 1988. Two couples were in Site C3, they were playing cards together in the evening when one couple, Susan and Scott Adams left to brush their teeth at the public washroom which was nearby. They had to walk down a dark trail with only a single light glowing far ahead next to the washroom door. As they were walking, they saw this thing standing on the trail between them and the washroom.

It made a large, deep-throated grunting noise at them. Scott said it reminded him of a large bull letting an immense amount of air as if to say, "I am here, don't come any closer". Susan initially thought it was a bear and ran back to the campground screaming, "there's a _ _ bear!". The other couple heard "bear" and ran to the cars with Susan. Scott slowly backed up and kept his eyes on it, he said he saw a large outline of the figure veer off to the left of the trail. When he finally got back to the camp and his car, his wife, Susan was so frightened she had locked the doors and wouldn't let him in. He did finally get in, and then the couples turn on the headlights of each of their cars. They see this thing walking right to left in front of them illuminated in the headlights. They all watched as this thing seemed intent on getting out of the area and disappeared into the trees heading towards Blackstone Brook.

They realized that they saw a Sasquatch and ended up driving around looking for it, while doing so, they flagged down this pickup truck with 3 other people in it. It turned out that these people said they saw something strange and were also driving around in search of what they saw.

What makes this interesting is that these witnesses did the rare thing, they reported it to the park warden's office the next day. Park wardens in Canada are like the US Park Rangers. Warden Allen Dibb told me when I interviewed him separately that he did everything he could to convince these people it was a bear. They didn't give in and insisted on filing a written report which should be still on file with the park warden's office at the national park.

I interviewed the four witnesses separately and their stories were really close. My only question is, who were the other people in the pickup truck? Were these 3 people trying to pull off a hoax and stuck around to see if they were successful? You must consider that. I was never able to track them down or find out who they were.

What excites you about this subject? Most of all what intrigues me is that the witnesses really believe they saw something. They are not lying or making it up, now whether they were mistaken or not, that is always a possibility. The footprints also, I have seen footprints at least 5 or 6 times that I don't think could have been faked. The best one and another favorite case of mine was the Bald Beard incident in 1986. I found 112 tracks across the road from the area where this American couple were on a fishing vacation. They said this big "gorilla thing" stole their fish! I caught them as they were leaving, they told me the whole story and later I found the tracks, they had no idea they were there. I cast two of them, the only ones that were clear were down by the riverbank. At one point under some mature trees, I found a cluster of tracks, it almost seemed like whatever made these tracks was walking all over the place like a man looking for his keys. I lost the trail at the edge of a rock-side area. I searched for three days and could never find where the tracks came out. I also never found the fish stringer that the couple says the Sasquatch took off a tree at their campsite.

If this event did not happen hadn't happened, I may have given up on the Sasquatch.

If Bigfoot is a large primate, what environmental conditions would it need to survive? The conditions which exist in the wilderness area of the Pacific Northwest today. If there is wilderness which can support a population of black bears there could be a Sasquatch.

How would you explain the elusiveness of Bigfoot? It has been the key to their survival. Stay in the shadows, and yield the ground to intruders. Avoiding injury is the key to survival in mountain country. Sasquatch has mastered this.

What do you feel you learned from the early researchers that helped you? I am lucky, I feel sorry for someone trying to get into this field today because there is so much junk that just gets dropped on your head. I was able to work with and get to know the pioneers. I learned everything drop by drop rather than having everything dumped on me.

Today people make big deals over twisted trees and such, I remember when the tree thing started, it was a hypothesis put forward by Bob Titmus, it was never an established fact. He was just wondering if Sasquatch had something to do with these broken-off trees we were finding, it was a question. By the time, the late 80s rolled around, every tv documentary was saying that twisted, broken trees were signs left by Sasquatch like it was a recognized truth.

The same holds true for these shelters and nests that people <u>want</u> to be Sasquatch's so badly that they throw all common sense out the window. A researcher may find out in the end that they may be wrong.

How do you feel that research has progressed in the recent years? We are still spinning our wheels. There are many more people now and they are doing the same things over and over again that we did early on, they just have prettier toys to do it with. I used to say we are not as bad as the UFO field although unfortunately the way it is now, it is. 90% are just nuts, which is a polite way of putting it. I call it "the asylum being run by the inmates", people believe that Sasquatches come to us through wormholes or the fourth dimension or UFOs or such. They try to use a mystery to explain another mystery, and I believe half of these people don't want to see the mystery solved at all, they like it the way it is. It is like an endless merry-go-round.

If we did not have the Patterson-Gimlin film, where do you think we would be now? Probably the same place we are now, The Patterson-Gimlin film seems to be a great piece of evidence, although there is a lot of other evidence out there. If it was proven to be a fake, my belief in the Sasquatch would go from about 90% down to 70%. There is so much more that I have seen and investigated. I believe the PG Film is real.

What other documented evidence do you find compelling? In 2013 the Port Renfrew video was filmed on Vancouver Island; you can't tell what it is and the couple did not know what they were filming. That one fascinates me. The video of a Skunk Ape in the Everglades that was apparently going after a snake is interesting. Also, I still think there is something to the Redwood tape taken in 1995 in Jedediah State Park, also known as the *Playmate Footage.*

What do you think about the Paul Freeman evidence? I liked Paul Freeman, I knew him, as a human being you couldn't have asked for better. I don't know if every track he found was faked, I feel he suffered from "Ivan Marx Syndrome", he was a classic example of it. A lot of people believe his second video was authentic, some forget he had an earlier one. I believe it was his son in a suit, possibly the same suit in some photos taken years prior. I feel he was trying to stay the center of attention and keep the interest going. I can't prove that, although it is my opinion.

Why do you think there is not more of the type of evidence to prove the existence of Bigfoot? I think it is elusiveness by nature. I feel Dr. Grover Krantz was correct by stating that he thinks there may be possibly 1 Bigfoot for every 100 bears. This is just a hypothesis on my part. If they exist, I think they live in family groups, one dominant male maybe 2 or 3 females and young. When the male offspring get to a certain age, the dominant male drives them off. The young males become nomadic and get bigger and stronger, and that is what people are seeing most of the time, these nomadic males wandering around. Then maybe one day, the nomadic younger males will find a family group with an aging male that they can either kill or drive off and take over. That is what makes sense to me, similar to orangutans and gorillas.

What is your elevator pitch for open-minded skeptics? I would say do your research, I gave up trying to prove this to the world two decades ago, I do this for myself. The only way we will prove it is when science gets what it demands and that is a body or a piece of body or sufficient skeletal remains.

What suggestions do you have for researchers who want to get involved in this field? My philosophy has always been, to stick to the facts and never deviate from the facts. Keep an open mind although stick to the facts and don't follow any agendas. There is one golden rule for anyone who wants to get into research, "Thou shall not hoax!". In addition, remember, we are not supposed to be "religious leaders" pushing a faith.

How to keep up with Thomas Steenburg:
ThomasSteenburg.com
e-mail: Sasquatch@Telus.net

DIANE STOCKING

Diane Stocking is retired from the Oregon Department of Human Services and holds a Degree in Forestry from Lake City Forestry College, Florida. She has been researching Bigfoot since 1974, lectures at various conferences and has made numerous TV appearances. In 2007, she co-authored the cryptozoological book *Elementum Bestia*. She founded the Stocking Hominid Research Group, creating a resource scientifically study the species and an avenue for witnesses to report encounters.

Interviewing Diane was an amazing experience, she is so much fun and totally candid! She is a wealth of knowledge and has decades of experience on the subject of Bigfoot. Not only is she one of the few women involved in the subject, but there are also only a small number of researchers that have been studying this as long as Diane has. She gets in the woods, interviews witnesses, vets out hoaxes and misidentifications, and is well respected by her peers and enthusiasts. It is evident that Diane is not involved with researching Bigfoot for self-promotion reasons, her only focus is the discovery and protection of the species.

Where did you grow up? South Miami, Florida. A little place called Cutler Ridge.

What are your hobbies? I enjoy getting out in the woods, hiking, and researching Bigfoot. Of course, grandchildren, they are important too. I have five now so the boys have understood that they are not as important anymore, it is all about the grandkids. Just like the old saying, "Spoil them rotten and give them back!"

What type of music do you like? I listen to the old stuff; Bob Seger, AC/DC, Eagles, Styx, Fleetwood Mac, Joe Cocker, Meatloaf, who sadly just passed.

What is your favorite place you have visited and why? Roaring River. This is a very remote area that is off the Clackamas River in Oregon. Remote and beautiful.

If you could take one person past or present on an expedition, who would it be? Kathy Strain, she is a tough, smart woman. She would have my back as I would have hers. She will not freak out while out in the woods.

When and why did you get involved in Bigfoot research? Growing up just outside the Everglades back then, there was not a whole lot of civilization on the west side of US 1, mostly farms, tomato fields, beans, and then the Everglades. When I was fourteen, there was a sighting on Alligator Alley, a gentleman had hit something that he said was a giant gorilla. He was on his way to work at approximately four in the morning. At the time, Alligator Alley was a two-lane road going through the Everglades. After he struck the animal, he drove to a fish camp and called the highway patrol. After the highway patrol officer arrived, the witness, his last name was Smith, took the officer out to where he hit *something*. Next, the officer went off into the woods to investigate and see if he could find the injured animal. Suddenly, this *thing* stood up in front of him and startled the officer. He turned around, took off, rushed back to his car, and called for backup. For three days they looked for this thing; helicopters, dogs, you name it, and they never found it. It did put a huge dent in Mr. Smith's vehicle, which was Cadillac or Lincoln, back in the early seventies those were built like tanks and whatever he hit, left a huge dent.

You know, when I think back now, there must have been evidence, hair, blood, or something, but nobody got any, a missed opportunity. That is what got me into research, I started writing people, going to libraries, and getting every book I could find on the subject.

If they exist, what do you feel Bigfoot/Sasquatch are and why? I am of the opinion that Sasquatch/Bigfoot are of the same Genus as Homo Sapiens. Same or similar skeletal structure, an upright bipedal form of locomotion, and intelligent.

How would you explain the elusiveness of Bigfoot? Their intelligence. They have absolutely no desire to interact with humans.

What is the biggest takeaway from interviewing witnesses? I am always skeptical, so, as it is great to get sightings, normally they are old. We can investigate the area, but nothing is going to be there. Even if we follow up on a report that is only a day old, it is unlikely we will find anything. It is helpful though, for data collection, we can document the area, moon phase, time of year, time of day/night, etc.....

Is there a compelling investigation that stands out to you? Yes, when I lived in Florida, Tom Steenburg contacted me about a sighting report he received. He received a letter from a woman who worked at a water treatment plant in Titusville, Florida. Since I lived in Titusville, Tom thought I may want to investigate the encounter. Two workers on the night shift had a visual encounter and they claimed that the Bigfoot left handprints on a window. I arrived and examined the handprints, they were compelling, I could see imprints of the hair around the wrist of whatever left them. They were extremely oily as well, I swabbed them and sent samples off to Todd Disotell, a professor of Anthropology at NYU who studies DNA and genetics in primates. Unfortunately, the summer sun in Florida baking down on the window degraded the evidence and he could not extract anything conclusive.

I also lifted the prints off the window and sent those off to law enforcement forensic expert, Jimmy Chilcutt. Like the other results, the only thing that he could determine was that the handprints were of primate origin. These were the most compelling pieces of evidence I have found. Twenty years later, although faded, I still have the handprints.

When it comes to footprints, the largest ones I found were in the Green Swamp in Florida. They were nineteen and a half inches long, they were huge. I found them in an old grapefruit grove and I mean old because these grapefruit trees were enormous. So apparently, somebody had planted them decades ago and then abandoned the site, but they kept growing. So that is where I found those prints. It seemed Bigfoot liked grapefruits, they were eating them. I think animals in general are smart, they adapt and work with what they have in the environment.

Handprint Images

What excites you about the subject? Oh, showing, describing, and proving that they are around, is not what excites me, it is what comes after that, learning about them. It is not going to be easy; we are trying to find something that is elusive, that is intelligent, that moves. So just understanding basic animal behavior is not going to cut it. There is so much we do not know about them, their gestation period, lifespan, are they nomadic or are they staying in a certain range all their lives? Those are the types of things that excite me and keep me going.

What is the strangest thing you encountered while researching? One of the weirdest things that happened to me, I was checking out a sighting in the panhandle of Florida. These two teenagers were out hunting and they saw one. So, I went out to the area and investigated. It was the weirdest thing because I was walking down this fire lane and suddenly, I got this feeling to turn around and go back. I had never experienced that before. Was it a type of infrasound? Who knows? It was a strong get-out-of-here feeling so. I turned around and went back to my truck and left.

How does your forestry background help your research? I have always loved the woods, the quietness, and not being afraid while out there. In Forestry we also studied animal behavior. This helped to better understand the characteristics of the animals that are native to each environment and maybe will help with studying Bigfoot if I ever find one.

What do you think about possible evidence that is shared on social media? I try to tell people, do not believe everything you see; many people out there *want* to be able to see something so bad they look at a picture and think it must be a Bigfoot. I disagree often if the evidence is not there, I have always been diligent to identify hoaxes and misidentifications. I have received countless pictures of a rock or a tree, and some of these may look like a Bigfoot sitting on the ground although they are not one. I tell these people to walk up to it, see what it is, the typical response is, "No, I can't do that."

I am not skeptical as far as whether they exist or not, I am of the opinion based on evidence that I have seen that they do. I am skeptical of all the crap that is thrown out there, "This is a Bigfoot, that is a Bigfoot." People need to look closer at what they have and pay attention to all the other garbage that gets put out there. Many people are trying to get attention, which is frustrating. Some of us are not out there for glory and to make a name for ourselves. I am not one to promote myself, that is for sure, I am the grunt, I am the one out in the woods and do not care about any kind of notoriety or fame. I just want to be able to find evidence so that it could be proven, that is what is important to me.

Who were the early women researchers? Back in the day when I started doing research, there were not many women involved, Barbara Wasson, myself, Bobby Short, Kathy Strain and not many more.

Did you research with Bobby Short? Yes, and we collaborated often although we did not go out in the woods together. She was a good friend of mine, we argued a lot, and I did not agree with a lot of theories that she put out there and she was the same with me. She was a smart woman.

What can you share about Barbara and Kathy? I didn't know Barbara very well. She had an arrogance about her. Kathy, on the other hand, is open to ideas but doesn't accept anything without facts and evidence. She's very intelligent. Kathy has an analytical mind and she doesn't take crap from anyone. Kathy is definitely someone you can learn from if you are willing to listen.

Why isn't there more of the type of evidence that would help prove the existence of Bigfoot? They are intelligent, they are problem-solving, like ravens, they do not want to be around us and they move. They are constantly moving and they do not want to be seen, that is what makes it so hard. They know their environment.

Do you think they live in groups or tribes? I think more like family groups, a clan, where they hang together, mom, dad, grandma, grandpa, kids, and grandkids. Like any other primate, the young males will probably leave and will go find their own little group or find a female and start one of their own. Researchers should study and learn primate behavior; it is important to understand.

Based on the reports you have taken in Florida versus the ones in the Pacific Northwest, do any of the physical characteristics of Bigfoot differ? They are mostly the same, they may have different hair colors, or longer hair, like other primates. We have different hair color, although our skeletal structure and morphology is all the same. I did notice that there are more reports of the reddish and chestnut color Bigfoot in the South versus in the Pacific Northwest, where we get more reports of darker hair color. This color difference could be attributed to the sun; it happens to us.

What do you suggest for new researchers? Go out in the woods, hike, learn animal tracks, learn about the environment, the food sources, things like that. So, you know what you are looking at when you are out there. When you find something, use critical thinking to figure out what you are looking at. Eliminate all other possibilities before you jump to "it's Bigfoot related".

What advice would you give women who want to get involved in Bigfoot research? Grow a tough hide and do not take any crap. It is harder for women, especially those with children and a job. They do not have as much time to do field research. Unfortunately, men are still seen as having more credibility than women. That is so wrong. I really feel women have a better chance of interacting with Sasquatch. Look at the greats.... Jane Goodall, Dian Fossey, and Birute Galdikas-Brindamour.

BOB STRAIN

Bob Strain was an Emergency Medical Technician (EMT), a Paramedic and/or Emergency Medical Services (EMS) provider/educator for more than 30 years. He was an EMT instructor for 12 years and a Firefighter. Although he had a personal encounter with a Bigfoot while hunting in Idaho at the age of 18, he did not get interested in the subject until many years later. After reading a few reports, Bob noticed that there were similarities in those reports and what he had experienced in the past which then began his journey into researching Bigfoot. Through his interest, he met his wife Kathy and Bob gleamed that this was the best thing that has come out of his interest in Bigfoot.

Interviewing him was an absolutely fun experience. There is no "BS" with Bob Strain and as he may seem brash and blunt, don't let that fool you, he is funny, compassionate, and extremely easy to have a conversation with. After discussing Bigfoot, I could have sat around a campfire and chatted for hours with him. Bob is experienced in the outdoors and has "street" smarts which certainly must help when interviewing witnesses and discussing the topic of Bigfoot. His experiences bring him behind the curtains of this phenomenon and allow him to communicate about the subject from a personal perspective.

Bob has been involved in several research groups over the years and has conducted outdoor and evidence collection training for researchers who want to get in the woods. Since 2012 he has been an active member of the North American Wood Ape Conservancy (NAWAC), taking part in multiple explorations in their research areas. Bob's diligence has resulted in multiple fascinating incidents which have helped the NAWAC team collect critical evidence and data.

Where did you grow up? I was born and raised in West Texas. When I was in junior high, we moved to Dallas.

Are you a Dallas Cowboys fan? You don't even have to ask that question; I don't have a choice. I got the shot, you know?

We may have to stop the interview [laughing]. As much as I am a Cowboys fan, there's no way you can hate them as much as I do. Do you know what I mean? You don't like them; I despise them sometimes. I'll root for them, but I want to kill them. [laughing]

What was your favorite concert you attended? I probably still have some hearing loss from it, it was Deep Purple when they played in Fort Worth in 1974 at the Tarrant County Civic Center, oh my, it was epic.

If you could take one person past or present on an expedition with you, who would it be and why? Daniel Boone, I imagine he would know everything about them.

How did you get involved in Bigfoot? The very first sighting I had in Idaho. I was hunting and watched one for several minutes through my rifle scope and binoculars. At first, I thought it was a bear. It was over on the hillside across from me where my dad was supposed to be. He was supposed to be trying to flush deer out into the open so I might have a shot. It was black and kept looking over its shoulder, I thought it was waiting on its cubs. It was on all fours and then turned and moved directly up this steep, loose rock slope. I'm thinking, if I shoot it, it's going to roll all the way down to the bottom and I thought I would wait for it to get up to the top and we could retrieve it better from there. Once it got to the top and I'm getting ready to pull the trigger, it stood up on two legs, looked around and started walking off, kind of like Patty does in the Patterson/Gimlin film. I then thought it was a man. This was 1975, and in Idaho, so for all I knew, it was some survivalist or a hippie living off the land.

I looked closer and I didn't see a backpack or a rifle. It was one solid color from head to toe and I could see the hands and the fingers. After I lost sight of it, I waited another 10 minutes or so. Then I see something orange way over to the right, in the same general area from which it first emerged from the trees. At first, I thought it was maybe a butterfly because it was so small. I thought, "What the heck is that?" I zoomed in on it and realized it was my dad wearing an orange vest. He was about half the size of what I had previously seen.

If they exist, what do you feel Bigfoot/Sasquatch are and why? The ones I've seen seem to be some sort of pretty smart ape that can walk upright or on all fours, climb trees and run like the wind.

How would you explain the elusiveness of Bigfoot? Bigfoot eludes humans because they can. Their great curiosity and forest ninja skills combined with incredible strength and speed result in a creature capable of physical feats outside the norm of human experience. Our minds often have great difficulty in separating perception from reality in split seconds.

Can you share a compelling witness account? Two ladies had hiked up the backside of a mountain, in Nevada heading towards the east side of Lake Tahoe. They were a mile or so up the trail when two or three teenage boys came running down the hill and zoomed right past them. They yelled, "Get out of here, there's a monster, there's a monster". The ladies were young and they just giggled and they kept going. Shortly afterward, they arrived at an area where the landscape leveled off and crossed the creek. One of them thought she had stepped in dog poop because there was such a horrible stink. They then looked up the hillside and saw - approximately ten feet away, a Sasquatch which was standing there looking at them. It was reddish-tan in color and had a hold of a small tree with one hand.

It pursed its wrinkled lips and hooted at them. One witness said it was so close she could see the cracks in its lips and that the skin reminded her of an old baseball glove. They hollered at it and it took off up the hill and knocked several trees away as it ran. They took off running down the hill and it followed them almost all the way to the bottom. It watched them from a rocky overlook as they got into their car.

What challenges do investigators face when taking reports? I'll be honest, you have got to be careful when you call people to follow up on reports. In the old days, it was easier, you could block your phone number, but now the person you call now has your phone number and suddenly now they're your friend. Every time they hear a whistle, you get a call.

We had one lady, bless her heart, I felt so bad for her. She called up Kathy in the middle of the night and whispered on the phone, "I have one outside my window, what do I do?". We had previously invested a lot of time with her and she did live in such a prime area, you would expect Sasquatch to be there. She rented the house and lived there by herself, and she was lonely. We figured that she hoaxed to try to get us to come up there and visit her, it was heartbreaking.

What are your thoughts on the Paul Freeman evidence? I think Paul got a bad rap because he was caught making things a little easier on himself. I think like this lady I just told you about, the reason we were up there originally was because she said she had a sighting and we found footprints, and there was an old, abandoned orchard. She was living in the mountains on the edge of civilization at 4000 feet elevation and at the edge of a lake. So, Paul Freeman, he was also in a great spot, you would have expected to find evidence if you looked long enough, and he did.

I think personally, the video that you see of the big one that's going through the evergreens is totally legitimate and the tracks that he found just prior to that were as well. If you look at the whole video, you can see that he's following the tracks and one of the tracks has water that is still filling in part of it. I think that's legitimate, but anything else associated with him, I'm just not well enough versed to weigh in on it.

What skills have you developed that help when interviewing witnesses? In the world of Bigfooting, you'll learn a lot more about human behavior than you will ever learn about Sasquatch behavior. I think what gives me a little bit of an edge, is my many years as a paramedic. I worked the streets, the ghettos, night shifts and I had heard everybody's line, you name it, I had probably seen it, I had been told every lie. One guy shot himself and told me that he didn't do it. I see the gun and I ask, "Why did you shoot yourself?". He says, "Well, are you going to give me some morphine?" I said, "No, not until you tell me why you shot yourself". He says, "I shot myself to get some morphine. Are you going to give me some?". I said, "No, I didn't say I'd give it to you. You shot yourself in the stomach, next time shoot yourself in the foot".

I get a good sense of when I am being fed a line of crap and if someone is being genuine. You also need a filter and I feel I have a really good bulls**t filter.

What type of research are you currently involved in? I would ultimately like to see the animal proven by science to be real and strive to conserve both them and their habitat. To accomplish these goals, there is no alternative than to follow the standards set forth by science and by government regulation.

What is your elevator pitch on the existence of Bigfoot to skeptics? Elevator pitch? I'm not familiar with the term elevator pitch.

If you met somebody in an elevator and had a few sentences to share why they should consider the possible existence of Bigfoot. Oh, I see. The time you would spend in an elevator. I thought you meant, like, flip them up in the air or something, because that's probably what I would do [laughing].

I've seen something that I can't explain and it fits no other description other than a Sasquatch. I've also seen enough evidence otherwise to convince me that Sasquatch is real.

What advice do you have for new researchers? If I were to give advice to someone looking to get more involved, I would go to a Bigfoot conference, make some friends, and see if there are any groups in your area. If there aren't any local groups, find a friend with common interests and wishes to go out and practice the things that you learn by reading and listening. Start out slow, the remarkable thing about Bigfooting is that it really doesn't take much to get involved. You just have to put yourself in the right spot. Go camping, many interactions that I've had have been within the area of a campsite. Going out and walking around can possibly elicit interaction, but just find a remote location, and be non-threatening.

I wouldn't suggest going out and doing some of these crazy things that some individuals do, like walking the roads at night in the dark. Do you think Sasquatch can't see just because it's dark? How do they run through the forest at night? They can see you in the dark, the only one that you're fooling is yourself.

When researching, be prepared for evidence collection, take safety precautions, learn the environment, and most importantly enjoy being out there.

KATHY STRAIN

Kathy Strain has been interested in Bigfoot since she was a young child. Professionally, she has been the Forest Archaeologist and Tribal Relations Program Manager for the Stanislaus National Forest for the past 23 years. Prior to that, she earned a B.A. and M.A. in Anthropology. She has also authored the book, *Giants, Cannibals & Monsters: Bigfoot in Native Culture*, which is a collection of stories pertaining to creatures in North America. These accounts have been passed down for thousands of years by the elders and tribal leaders from 55 Native American cultures. She had her first Bigfoot sighting in 2012 where she along with 4 others witnessed 2 of the creatures.

Kathy has a refreshing personality, intelligent and scientific yet light-hearted and fun-our conversations have always been enjoyable. It is evident that family and close friends are a huge part of her life, she lights up when discussing them. I certainly enjoyed hearing how

Kathy met her husband Bob and learning of some of their adventures since. Speaking with others who are involved with her peers, I can see that she is well respected and has developed many lifelong friendships.

Kathy is a current member of the North American Wood Ape Conservancy (NAWAC), previously an investigator for the Bigfoot Field Researchers Organization (BFRO) and has provided professional input for multiple television and movie documentaries regarding the subject. You may also find her speaking at a conference or educating enthusiasts on field research and history. Kathy continues to be part of a small group of academia that publicly studies the Bigfoot Phenomena, shedding hope that science may soon invest more attention to the subject.

What are your hobbies? I like to garden and watch old-time movies with Bob, that's always fun, Alfred Hitchcock and Perry Mason, that kind of stuff. Art Nouveau, I think is what they call it.

What are your favorite musicians or bands or type of music? I'm definitely an 80s kid, my favorite band is the Cars and I really like Kiss.

What's your favorite childhood memory? You're talking to an old lady who's got lots of memories [laughing]. I was very fond of my Grandma Rachel on my dad's side and wishing now that I had spent more time listening. I was never quite the same after she passed away. My mother has now also passed and my favorite memory of her is sitting behind her recliner as a child to watch *Legend of Boggy Creek.* I figured if the monster came through the TV, she would save me.

If you could take one person past or present on an expedition with you, who would it be and why? Jesus Christ because he would know where all the Bigfoot are!

If they exist, what do you feel Bigfoot are and why? I feel that they are primates, specifically an unknown ape. Scientifically, since no other animal walks on two legs except us primates, they would fall somewhere in our evolutionary history.

How would you explain the elusiveness of Bigfoot? I think they have learned that for their species, staying away from us keeps them safe. I think Native Americans in prehistory likely saw them more often than we do in modern times because there was a larger population of Bigfoots. Since we came along, I believe their numbers have dwindled due to our use of the environment (cutting trees, dams, pavement, etc.), which has resulted in them being seen less and likely their avoidance of us.

When and why did you get involved in Bigfoot research? I come about Bigfooting, differently than many because I have believed in Bigfoot longer than I've studied them. When I was a little girl, I saw *Legend of Boggy Creek* on television, which was unusual because my mom never let us watch scary things. I sat the entire time behind her recliner and watched because it really scared me. It was the kind of scared that intrigues you though. Right at that minute I just knew that this was the best thing ever. I really wanted to know more about what Bigfoot are and how I could study them. Every time it came on television, I made everybody watch. Then, sometime around fifth grade, I can't remember exactly, I asked my teacher what I had to do to study Bigfoot for a living. She said, "Well, you'll probably have to go into anthropology". So, I went into anthropology with the mindset that this was going to be a career, and of course, I soon figured out that nobody was going to pay me to study Bigfoot.

However, anthropology and archaeology were a natural fit as a career anyway. My family always took three to four weeks each summer to travel and we always went to national parks, forests and historic sites. We always made our dad stop at the roadside historical markers. We rarely went to places like Disneyland, that just wasn't who we were as a family. I also loved the outdoors already, so archaeology was a natural fit.

I don't think I saw the Patterson Gimlin film until it was on that Leonard Nemoy special. My world was more focused on following researchers like John Green, and that kind of level. Of course, after I saw the Patterson Gilliam film, I said, "Whoa!". I still remember watching it and running and getting my mom and saying, "Mom, you have to watch this". She was also impressed. When I look back, in many ways, I must give my mom a lot of credit because she was always encouraging and she was just as interested in Bigfoot as I was. She bought my first John Green books for me.

I was lucky that I started my career in the Sequoia National Forest. I was able to work with the local tribe that have the *Hairy Man Pictographs* on their reservation. The tribe members told me many of their traditional stories, and it was then I figured out that Native Americans had a much deeper, longer history with Bigfoot, more than I could comprehend at the time. Most people at that time thought Bigfoot started with the Patterson Gimlin film and prior to that, they didn't exist, which we all know is not true. Previously, I had visited the site of the *Hairy Man Pictographs,* although I didn't understand what they were and their significance until later when I worked with the tribal elders to better understand their meaning and importance. Their views impressed me.

That's how I came into Bigfoot, it's the complete opposite of where most people have sightings and then they go into Bigfoot. I didn't even see a Bigfoot until 2012.

What are the *Hairy Man Pictographs*? The *Hairy Man Pictographs* are prehistoric paintings with contemporary stories told by the Tule River Indians about *Hairy Man.* Located on the Tule River Indian Reservation, the pictographs are approximately 1000 years old. According to members of the tribe, the pictographs depict how various animals, including *Hairy Man*, created People. Other stories tell why Hairy Man lives in the mountains, steals food, and still occupies parts of the reservation. Since the Tule River Tribe equates *Hairy Man* to Bigfoot, the pictograph and stories are valuable to our understanding of the modern idea of a hair-covered giant.

Was your experience in 2012 you're most compelling one? Oh, it was, it was a dream come true, it was very compelling. I have had plenty of other things happen to me in my life, but nothing to match that under any circumstance.

We had gone to Area X on the request of Alton Higgins and Daryl Colyer. They wanted our opinion on what was happening down there. Bob's family lives in Dallas, so we were just going to make it a family adventure. The first day we got there it was typical Oklahoma in the summer, always trying to kill you at all fronts. It's not a fun place to be, it's hot, sticky, and humid. Essentially there are four cabins down there, all with tin roofs.

The first day was beyond boring, outside the good company and friends, there was just literally absolutely nothing going on. The next day everything changed. We spent the afternoon chasing something that was throwing rocks on the tin roofs. We would run over to one cabin after we heard a rock throw and then there would be a rock throw at the cabin we just came from, it was like a game. We were running back and forth and I eventually decided I wasn't going to do that anymore so I stayed behind while the rest of the guys (Bob, Brian Brown, Mark McClurkin and Ken Stewart) went to the other cabin. Some time passes and I'm standing on the back porch and I see a branch move in a tree approximately 30 feet to 40 feet from me. It gently released like somebody had pressed down with a hand to peek at me and then realized I was looking at them and let it slowly come back in a controlled movement.

I was thinking, "Oh, I wonder what the hell that was?". The guys came back and I said, "Hey, I want you guys to go over and look to see what is in that tree." Brian Brown is the one that went in first, he looked and said, "Oh, there's nothing there, all I can see is a footpath where animals have been walking." Then there was another rock throw and so they all took off again to investigate. I asked Bob to stay behind and asked, "Can you really get a look in there? I got a really creepy feeling that I was being watched.": There wasn't enough time between when I saw it and when the guys came back for whatever it was to have really left the area. So, he walks over and he says, "Is this the location?" and I responded, "Yes", and he sticks his head in, and he looks and said, "Oh, honey, there's nothing in here but a couple of down logs."

A little later, the guys came back and we sat outside and were chatting. I'm looking down the bottleneck, an old driveway that's grown back, it's mostly grasses versus most of the rest of the property where you're dodging limbs, thick trees, or getting green brier-patch cut up in your shoes, it's an easy walking location. Mark McClurkin says, "What's that sound, it sounds like something's walking?" We all heard it and thought it was this fox that we had been feeding because he was super cute. I'm looking in the direction where the sounds are coming from and all of a sudden there's two Bigfoot coming straight at us from the same location where I saw the tree movement.

They were walking along the edge of that mountainside and heading towards the north cabin. I think they were trying to get behind the shed that's there, but I am not certain. It looked a little funny because there was a big one and a little one, and it looked like the big one was trying to corral the little one and get him under control while the little one was just going crazy. It was like chasing a six-year-old who's got his mind set on running to the candy machine or something. I jumped up and I yelled. "There they are", and I ran at them. Bob, Brian, and Mark saw the same thing I did. Ken Stewart is always looking somewhere he's not supposed to and so he didn't see anything. He's seen more people see Bigfoot than anybody because he's always doing something other than looking for a Bigfoot.

I run at them and they turn and bolt up that hillside like nothing you've ever seen before. It looked like they were on a bungee and got released. They were absolutely silent running up that hillside. I focused on the big one because I figured that is all I had time for. I could see its backside, its thighs were as powerful and muscular as anything I'd ever seen in my entire life, the buttocks were muscular. It had tucked its arms into its side, like a power runner. I noticed the big one never looked back to see where the little one was, so it couldn't have cared less, which was funny. I always had the impression that it was an older sister watching a younger, bratty brother, it wasn't her kid so she didn't care. He was to the right of her, and it was clear that he was moving just as fast. He also put distance between them, almost like they knew, that if we were going to run after them, we'd have to split up. Obviously, they were trained, they knew how to elude an enemy.

Everybody ran behind me and saw the same thing. We watched it as long as we could, we were not going to be able to catch them, it would have taken us at least ten minutes just to find a cut path to contour and get up on the hill.

The next morning I'm the last one up and I come out onto the porch after getting dressed. I opened the door and it just felt like the heaviness, like there was something wrong. I look over at Brian and Bob, but Bob wouldn't look at me. Brian says, "Do you remember what Bob told you yesterday when he went over there and looked in that bush?" I said, "Yes, he said, honey, there's nothing here, but a couple of downed logs". Then Brian states, "Guess how many logs are there now?". My mouth hung open and I said, "Oh my God, they were there the entire time."; Essentially, they were playing dead, it's something that's hair covered and psychologically that makes complete sense to me because we see what we want to see. There was one log already in there, so it makes sense Bob thought he was seeing multiple ones. You're expecting to see an animal, not something that's pretending to lay still and be ambiguous.

The irony is that our mission at the NAWAC is to get a specimen and there we had the perfect chance to do that and we didn't. I also believe in fate; it wasn't the time or the place. What if Bob had stepped on one of them and it jumped up and killed him or something to that effect? There were so many variables and I believe in divine intervention, what happened was meant to happen. I still had an exciting time with it, I got my first sighting and I wouldn't trade it for anything.

Kathy Strain

When you chased them towards the hill, how long was the sighting? Maybe 10, 15 seconds, it was very quick. The worst part about it is we had just put a camera out and I'm the one who triggered it, so all you see is me running past the camera.

Scaling the hill was like child's play to them, even deer aren't that graceful. It's hard to explain, like a beautiful ice skater, just as skilled and quiet yet graceful but powerful at the same time because of the muscle depth. This wasn't even a big one, it was probably six-foot tall and I doubt fully grown, and the little one was approximately four feet. The bigger one had muscle structure, it was obviously born that way, just like gazelles, for example, a particular muscle structure for their environment.

Approximately how many eyewitnesses have you spoken to? Oh, hundreds. I used to be in the BFRO prior to the NAWAC. I've also talked to plenty of people who seemed to not be telling the truth as well. You just get a feel for if somebody's telling the truth or they aren't, I guess is the best way to put it.

How did you meet Bob? I used to be in the BFRO and I handled the vast majority of the sightings in Tuolumne County up through

Plaster, Nevada and Lake Tahoe. At that time, Bob was coming to grips with his own Bigfoot sighting and was just getting into Bigfooting. He was a huge, gigantic fan of mine [laughing]. I was speaking at the 2003 Bigfoot International Bigfoot Symposium, it was the first symposium I had ever attended. It was also the first time I spoke in person about the subject, I was outing myself, essentially. I used my name for the BFRO, but I hadn't really come out as a Bigfooter publicly or with my peers of archaeologists and such. Many of the individuals I met there later became dear friends, Alton Higgins, Daryl Colyer, Tim Cullen, and Craig Woolheater, I could just go down the list. The speakers were also top-notch; John Green, Jeff Meldrum, Doug Hajiceck, and Bob Gimlin, this was the first time he had spoken in public since the film was taken.

The speakers had been gathering at the beginning of the event, I missed it because I was still exploring the museum. Bob had been getting the speakers to sign this T-shirt, and so I walked up to him and I asked, "Do you want me to sign it?", and he just walked off. I don't know what I did, but I assumed that this guy was unhappy. Afterwards at lunch, I am talking to Tim Cullen and Bob comes over and sits next to him. I say, "Hey, Tim, who's your friend?". I was just intrigued by him, and thought, "Wow that dude's really good-looking." I then got called away to handle something related to the event. As the event goes on, I continue to notice him and I never got to talk to him until the very last day. I was packing up my car and preparing for a tour down at the Bluff Creek site. Suddenly, Bob taps me on the back of my shoulder and says, "You know, I'm trying to get this T-shirt signed by all the speakers, do you mind signing this for me?". We walk back inside, Alton and others are there, Bob introduces himself and hands me the pen and the shirt. I don't even know why, and it was so not like me, but I said, "Would you like me to sign your chest instead?". Alton Higgins says, "Kathy, behave yourself", and I just thought I was so funny, I just laughed. Alton didn't think it was funny. Bob just had this look on his face expressing, "What am I to do with that?". I take the T-shirt and sign it. Later he ended up contacting me and that's how we met. It's a funny story. The funniest part is he missed my talk altogether.

2003 Symposium

What excites you about the subject? I think that the fact that we still have a mystery that humans haven't discovered. There is something out there that's still wild and untamed and has the luxury of not having a cell phone, bills to pay, or listening to politics all day long, I think that's a beautiful thing. It is very blessed. There is still something left undiscovered that isn't built on, or available at a cost, you can't purchase a Bigfoot, you have to actually go outside and look for him. I think that's what's exciting.

So how do you feel mainstream scientists will publicly accept the existence of Bigfoot? We have the PG film and they still don't buy it. We are going to have to have a body. If we don't get a body, I don't see science ever accepting that they exist. They're going to have to be able to touch it, analyze it, get DNA, blood type it, look at its hair, all those little things that you would expect for any other species. Hopefully, they don't need 10, just the one that we can share with science. We probably have DNA from them in other places and don't realize it because we don't have a typical model to compare to.

What's your elevator pitch for open-minded skeptics? Well for me, I am a professional archaeologist and I've had a sighting. Individuals don't meet many people that are scientific and have had this happen. For me, that's the biggest pitch, I'm clearly not crazy, not a liar, I've earned the respect of my peers, and yet here I am saying I had this experience. I saw two Bigfoot in daylight, there's no way I made a mistake, I've seen plenty of wild animals in my life. I've been a field-going archaeologist for 30-plus years. I feel it makes people think, "Whoa, well, wait a minute, maybe there is something more to this than I originally thought". At the same time, I do understand some of the reluctance in the sense of you have to see it to believe it yourself.

What suggestions do you have for new researchers who want to get involved in Bigfoot? I would start by learning who came before you. You can't be a Bigfooter and not know who John Green or Grover Krantz is, that's unacceptable. Know who came before you and what they did to make it possible for you to be here. You're not starting on nothing, you're starting with some level of the work from previous researchers, you're standing on the progress that they made. So, give them the credit to know who they are, start with that. Don't read the fluffy fictional crap. Read Jeff Meldrum, John Green, John Bindernagel, Grover Krantz, understand what you have available, and learn from them. I would stay away from the information that's just flat-out crazy, don't go down that path of the "woo" stuff, recognize it for what it is and just stay clear of it. Do ask questions though and feel free to share what you think has happened to you and don't be offended if people question what you're saying.

LES STROUD

It is exciting to have Les Stroud be part of the book and thank you to Doug Hajicek for the introduction. Most of the world knows Les Stroud as the outdoor survivalist and star of the groundbreaking and award-winning hit tv documentary series, *Survivorman*. In addition to being the star, Les also creatively wrote, produced, and filmed the series. Since the concept of *Survivorman* was based on Les filming himself alone in the wilderness, audiences got an up-close view of his life. In 2014 he produced a 9-part series titled, *Survivorman Bigfoot*, where Les explores the phenomenon. He leveraged his experience in the wilderness and also took an objective view of the Bigfoot subject.

Getting to know Les has been enjoyable as it has with the others in this book. I appreciate that he speaks openly and does not hold back on how he feels, he had no problems pushing back on a few of my questions. I will add, he is intently adamant about not being called a Bigfoot researcher, his many years in the wilderness and curiosity fueled his interest in the subject.

My conversation with Les was memorable and candid, we certainly touched on a multitude of topics related and not related to Bigfoot, some controversial and all honest. Since our initial correspondence, he has graciously been available when I needed additional information for this book. Les has positioned his success with *Survivorman* to provide a global platform to explore the Bigfoot phenomena for a broader audience making him important for the subject.

Where did you grow up? Toronto, Canada

Hobbies outside of BF? Adventure and travel. Music is incredibly important to me and more than a hobby, it's my other vocation. I've recorded six albums; I perform and tour. Fun hobbies are basically hockey and paintball.

Favorite musicians or genre? Very eclectic actually. Classic rock will always be a go-to for me because that's my era. But other than that, at home I might be putting on Miles Davis or John Coltrane, Dvorak, Mazursky, Sonny Boy Williamson, Sonny Terry, Brownie McGhee, Black Sabbath, Yes, or Genesis and of course classic rock. So really everything saving for most modern country or hip hop. Also, as a musician, I do try very hard to have an open mind and listen to other forms of music that are out there, whether it's country or hip hop or whatever, most of which I hate but still... I try. Blues has been a big part of my musical life as well and when I took harmonica seriously, it meant that I needed to take the blues seriously. So, I kind of entrenched myself in it for a long time.

Favorite place you have visited? I've been able to travel the world because of *Survivorman* so my three top beautiful places will always be the high Canadian Arctic, the Amazon jungle leading up to the high Andes and then the Utah Canyonlands.

If you could take one person past or present on an expedition with you, who would it be and why? I accidentally thought of this a while ago. I would take John Denver. I have always very much related to his lyrics. I've always wanted to connect people to nature and I feel like he did that with his music better than anyone that I've ever listened to when it comes to lyrical content. He is the only artist that really affects me emotionally. We lost him too soon, and I believe that if he'd been alive, he would have been a *Survivorman* fan and I would have loved to have just had the chance to take him on a canoe trip.

What have you learned about yourself from being out in the wild alone? I don't want to be overly dramatic about it, I certainly don't want to be cliché. It seems like the more significant my work is, the less significant I feel. I think if I've discovered anything, it's that I have an inherent need to make my life matter.

What's the strangest thing you have encountered in the woods? It would be Bigfoot or at least what I perceived to be Bigfoot anyway. And there's a collection of those experiences because of doing the series *Survivorman Bigfoot*. On the top of the mountain with Todd Standing, the feeling of getting sat on and seeing lights in the sky. The apples disappearing off the tree and the little head showing up at the end of the video. Then there's the ape-like grunting that happened to me in Alaska when I was filming *Survivorman*. There was also the screaming I heard as a younger man training in survival in southern Ontario, that was just satanic. In truth, there are several moments for me.

Was there ever a moment when you were afraid for your safety? I guess the toughest would be coming down off the mountain in the Norway episode of *Survivorman*. In that instance, I put myself in a precarious situation by getting cocky about my abilities. I was very close to becoming hypothermic. I was soaked to the bone with sweat and soaked to the skin with rain. So, I got pretty nervous and fearful that day. I did make it down to the bottom eventually. But man, that was a tough day for me big time because I came from 10 feet of snow to no snow and freezing rain all the way down. So it was pretty rough.

What is the first rule for someone who wants to spend time in the woods? Train, train, train. It's a good question because people think they can watch an episode of *Survivorman* or any of the copycat shows and then just head out and survive in the woods. But they miss the fact that I spent 15 years training and then instructing before I went off and filmed Survivorman. Let me put it this way; You wouldn't watch Olympic ski jumping on TV one day and the next day strap on a set of skis that you've never put on before and go to the top of a jump. You wouldn't do that, and it's the same thing with survival. You shouldn't just watch people out in the wilderness, never mind the reality bullshit shows, but people who are actually doing it for real out there and then go out and do it tomorrow yourself, even though you've never worn a pair of hiking boots in your life. So, train, train, train.

What is the funniest moment you have had in the woods? I'm so slow and boring out there that not a lot of dumb-ass things happen to me. I'm always taking a lot of care and concern because I'm truly in the middle of the Amazon jungle by myself and if I get bit by a Fer De Lance, that's it. So, I'm really slow and cautious and careful, notwithstanding the day in Norway. Which does not translate into a lot of comedic moments, to be honest.

I guess the funniest moment, however, would have been while filming the Labrador episode of *Survivorman*. Every single night I had snowmobilers from the local native village come out to get my autograph, they would have me write my name on the windshield of their snowmobiles. Meanwhile, I was trying to film. They were really, really nice people. They were just saying, "Hey, how's it going, eh?", they just wanted my autograph and to hang out. They were bringing me bottles of scotch and smoked trout, and I'm like, "Hey guys, I can't take it, you understand what I'm here to do right?", and they're like, "Oh, OK, well, can I get your autograph?". It was hilarious, man.

What compelled you to make *Survivorman Bigfoot*? It was sort of three parts equal to my own fascination with the subject, the knowledge that I'm a good documentary filmmaker, and a reaction to how poorly *Finding Bigfoot* had treated the subject matter. So to elaborate is to say that I prefer to work on creations that come directly out of my own interest. Non-derivative, that sort of thing. Every once in a while in life I think the Richard Branson version of "How to Succeed" is possible, which is the "make a better mousetrap" kind of philosophy. Sometimes I will say, "I can do that so much better". When it came to *Finding Bigfoot,* and I know you've heard me say this in multiple interviews by now, the problem I have wasn't the cast. I think Cliff is a really cool guy, I like him a lot, he's a wonderful man. I don't know any of the others, I don't know Matt Moneymaker.

But what I can say is about what that show did to the phenomenon of Bigfoot; it turned it into a cultural punchline. I then realized that now no one will ever take this seriously, it's just too funny for people now. For example, there's a reference to it in a Robert Downey Jr. movie where he's playing a lawyer and he says, "and my client wants to see a Sasquatch." I remember thinking, "Oh God, now it's just a big joke for everybody". People can't say Bigfoot without giggling. The antics of Bobo were laughable and inane and very unfortunate. So, I was a little.... well not *incensed*, but definitely irked thinking, man, "You guys took a phenomenal opportunity and you blew it". I'm not saying they did that intentionally, they made millions of dollars and big ratings, good for them. But they still ruined an opportunity to actually take a very interesting phenomenon seriously. After that, I thought, well let me see if I can do *something* to inch it backwards from a place of, "Hey, can we not all just be interested in this without being laughed at?". Then if you add in my own different experiences I had and the fact that I knew as *Survivorman*, I was perfectly positioned to get financed and do a proper documentary series, you end up with *Survivorman Bigfoot* being produced.

How do your possible encounters with Bigfoot differ from other animals? They're not in the same realm at all. With other animals, it feels like you're dealing predominantly on an instinctual basis. I won't say that there isn't anything larger than just instinct involved here, I'm sure there likely is. But for the most part, I will deal with a bear or a cougar or a wolf or a lynx or a fox or a squirrel *instinctively*. It's a lot of instinct and *energy*, and how your *energy* is and how their *energy* is. I saw that frequently with sharks, for example. But with Bigfoot, you must ask yourself, "Am I dealing with Gigantopithecus or not?" If it's just a big smart ape, well then maybe we're back to instinct again. However, I don't believe that it's a big smart ape, so then it means that we're dealing with such an unknown that instincts are not going to be enough. Your instincts might alert you to the fact that something strange is going on and something strange is there, but instincts are not going to be enough when it comes to an encounter with something known as Sasquatch. That's something else altogether.

If they exist, what do you feel Bigfoot are and why? Originally, I leaned towards Gigantopithecus simply based on the platform that I first learned about them. It was the work of Jeff Meldrum and his book, Doug Hajicek's films and those kinds of things. Much of it was the research into the Gigantopithecus jawbone and John Bindernagel and such, so that's where you start. That is sort of ground zero for you, and if you're trying to interest somebody else in this whole phenomenon, because you don't jump right in with portals and aliens. So, you have to start somewhere and that's where I started although I didn't have a preconceived notion, I thought, "I'm going to get into this now, OK, what do we have out there? Show me, what you've got?". That was it, I was an open book with an open mind to whatever was possible.

What excites you about this subject? It's one of our human existence's ultimate mysteries. What is out there that isn't human, and by the way, at the end of this, if that's real, what *else* is out there? So, it's an ultimate mystery, but it's one *deeply* associated with the natural world, with the wilderness, with nature. The overwhelming majority of any discussion around Sasquatch ends up with you talking about nature. I'm all about nature, I'm all about connecting people to the wilderness, to the natural world. And Sasquatch seems to be all about representing that connection to the natural world. So, it's a no-brainer that I'd be interested.

If Bigfoot is a large primate, what environmental conditions would it need to survive? The cop-out answer is "I don't freaking know!". But if we look at the attributes that are associated with the thousands of anecdotal references, we come up with the existence of *something*, a phenomenon maybe, that has both physical and non-physical manifestations. In its physical manifestations; apparently, it needs to eat, it needs to defecate, it stinks, it can swim, it can climb mountains, it can scream, it may be able to do infrasound like a lion, it may be able to do cloaking like an octopus. So, you have all these physical attributes given to it by thousands of anecdotal references. So you compile a list. Then you must look at that list and ask, "Well, what does that require?". Well, eating requires prey or gathering, but either way, it's got to eat something. Is it leaves and berries or is it deer and elk? Does it eat humans? I don't know. If it's flesh and body, does it need to be warm in the winter? Does it need to be dry when it rains? Does it avoid the rain? Does it need caves? You really have to compile all of the various anecdotal references about attributes and then say, "So it's a species who needs to get out of the rain, it needs to stay put in the cold, it needs to get cool if it gets too hot. So, let's look at all of those needs, compile them and place the answer to that question in the ecosystem that would suffice.

Todd Standing is a controversial figure amongst researchers, what is your perspective on him? Well, that's a big question. If I can swear, it's f**king Bigfoot we're talking about, so *every* person involved is controversial. If you want to look at it that way. So let's start with his footage. I don't give a crap what all these people who put themselves in the position of being video technical experts have to say. I'm a filmmaker with 30 years of experience and I've had that footage in my hands. So as far as I'm concerned, you can't prove or disprove it. I think other influencers you speak with are going to have different opinions, they are going to be very black and white about Todd Standing, then I think they do themselves and the whole thing a disservice. Here's the reality, and it comes with a big capital "I" on the word "**If**"; If Todd Standing's footage is real, then it is bar none the best footage in existence right now. That is a tough reality for a lot of people to swallow.

What this means is, if you accept that, you're saying it's better than Roger Patterson's. Now, remember, all of us have a ridiculous love affair with Roger Paterson's film footage, it established everything, it's the go-to, it's the real deal, and all the naysayers don't know what they're talking about blah blah blah. So along comes this footage from Todd Standing. There's an initial reluctance, simply because it's the world of the unbelievable, and so that's understandable.

So now you have to look at the character of Todd Standing. I have various opinions on Todd Standing. I would still go out in the woods with him tomorrow and he still invites me. I cannot tell you if that footage is real or isn't real, and I've had it in my hands. What I can tell you, is that when Todd Standing speaks to a crowd or on film, he is challenging to listen to, I'm being very generous here. When somebody is "challenging to listen to" then automatically anything they present you is bu**$hit in our mind. So right away, everybody is like, "I can't listen to that f**king guy, he is full of sh*t. I can't believe him, he is just doing it for the money, blah blah blah, his footage is bu**$hit.". Yet in fact, this is a guy who perhaps *because of* his particular personality, as challenging as it is, ends up out in the bush for a month straight without coming home. The guy is more boots on the ground than any of his detractors.

So, my opinion of Todd, personally speaking, doesn't matter. But writing him off is pointless and prejudiced. If I were to throw the personality baby out with the bathwater when it comes to Bigfoot researchers, there's almost nobody I'd want to call. Well except for a few people. One would be Stacy Brown, he's the real deal. Jeff Meldrum, love him. I definitely would call him, because he's a scientist. And Devon Massyn. Also, David Paulides and Doug Hyjicek. But for the most part, if we're going to start throwing babies out with the bathwater, then you're going to lose 80 percent of your list.

So regarding all those naysayers on Todd Standing, I *get* it, he's very difficult to be in the same room with, I *totally, totally* get it. I spent a lot of time out in the field with this man, endless hours. I climbed mountains with him. That's why I admit it, and I would say to him, "F**k Todd, shut up for f**k's sake,". I would say that to his face, and he would get it and laugh about it. But I'm not going to throw that baby out with the bathwater as far as I'm concerned. Or you'd have to do that with everybody on your list because it's goddamn Bigfoot researchers!

I was in a heated argument with the top brass at the Canadian Discovery Channel over *Survivorman Bigfoot*. The idiot executive had somewhere heard there was controversy about Todd Standing and was warning me not to use him and that I should rather use their guy instead. I was like:" Are you freaking kidding me!? We're talking about BIGFOOT researchers - and you want to argue credibility with me??? Do you even hear what you're saying!!?? In any event, I stood my ground and as a result, I lost the deal and sale of my series in Canada. All because I refused to get rid of Standing.

I want to establish and make something really clear here, that is different between me and this scenario and this interview and your book. I am not a Bigfoot researcher; I do not believe in Bigfoot and I have not seen Bigfoot. At the most, I've conjectured and I've put out some hypotheses. What I am is a documentary filmmaker who did a series on a very interesting phenomenon, that's it. So, I want to make sure that you make that distinction with me.

What else about possible Bigfoot footage is interesting? I've got a very close friend who is a Bigfoot researcher and who made, on purpose, a fake film and put it out there without saying anything about it just to see what would happen. *ThinkerThunker* exclaimed his praise and said this is the real deal. I think footage from 20 years ago, 10 years ago, 15 years ago, is more credible, because now it's *really* easy to fake. So what does Todd standing have there? Does he have *some* real footage that people didn't go crazy over, so he made *some* fake footage? I don't know. Is it all real? I don't know. Is it all fake? I don't know. But that's hard, man, everybody wants me to get the definitive clip and I'll tell you right now, the last thing I want to do is be the person who proves Bigfoot once and for all. It's the last thing I want to do because that would define the rest of my life. I would have nothing else in my life, except I'd be the guy who exposed Bigfoot for real, and I don't want that, never did.

What is your elevator pitch for open-minded skeptics? Here's what I used to say. First of all, don't ask me if I believe in Bigfoot. But if you ask me if I think it's plausible that out there in a vast area of wilderness, is it possible that a bipedal upright, walking species exists, with incredible attributes and has been able to remain undetected, that is also responsible for hundreds of thousands of anecdotal references, including rock-throwing, screaming, sightings, footprints, hair samples, defecation? I would say, "yeah, I think it's at least plausible". These anecdotes span the entire globe, involving hundreds of different cultures explaining the same phenomenon by different names, so at some point, you have to, even as a scientist, say, "Well, there's got to be *something*. It can't all be mass delusion". I'm open-minded enough to say, "Well, I don't think it is mass delusion". There, now we are on the second floor in the elevator, pitch is over.

How do you feel mainstream scientists will publicly accept the existence of Bigfoot/Sasquatch? They need a body and that's really unfortunate. Part of me wants to say, who cares what they think? As much as I have complete faith in mainstream science, I also have zero faith in mainstream science. It depends on the subject matter.

Les Stroud

How to follow Les Stroud:
YouTube: Survivorman Bigfoot
Podcast: Surviving Life With Les Stroud

CRAIG WOOLHEATER

Craig Woolheater co-founded the Texas Bigfoot Research Center in June 1999 and started hosting the Texas Bigfoot Conference in 2001 in the Piney Woods of Texas, in Jefferson. In February of 2018, the Mayor of Jefferson, Texas signed and issued a town proclamation establishing Jefferson as the Bigfoot Capital of Texas. His organization, the Texas Bigfoot Research Center, was issued permits by the National Park Service to conduct camera trap studies to capture credible images of Bigfoot in the Big Thicket National Preserve.

The organization also collected bear hair samples using hair snares in SE Oklahoma that Texas Parks & Wildlife received and compared them to bear hair samples that TPW collected to compare DNA to determine the origin of bears repopulating East Texas. The organization also placed camera traps in Sam Houston National Forest under permission from the US Forest Service. He has co-curated two Bigfoot museum exhibits: from February through August

2004 at The Brazos Valley Museum of Natural History in Bryan, Texas and from April through August 2006 at The University of Texas at San Antonio Institute of Texan Cultures, a Smithsonian Affiliate. This highly acclaimed exhibit was entitled *Bigfoot in Texas?* He created the blogsite *Cryptomundo* in 2005. He has organized and hosted the Fouke Monster Festival since 2019. In 2021, he organized and hosted the Texas Bigfoot Film Festival. Craig has recently moved to Jefferson and is working on his 20-year dream of opening a Texas Bigfoot Museum there.

Craig's interest in cryptozoology started early before he was 10 years of age. The case of the Lake Worth Monster in the summer of 1969 hit close to home for him as a young boy. His reading of the John Keel classic, "Strange Creatures from Time & Space" was the next piece, followed by a screening of "The Legend of Boggy Creek" when he was 13. His reality came crashing down on Memorial Day 1994 when he witnessed what he can only describe as a light-colored Bigfoot while driving in Central Louisiana late at night.

Craig is passionate about educating Bigfoot enthusiasts and one of his most impactful skill sets is the ability to successfully organize forums on the subject. Through his efforts over the past 22 years, he has provided witnesses, enthusiasts, and researchers in Texas and throughout North America with an avenue to communicate and collaborate. Equally important, because of Craig, groups, and individuals who have not typically been interested in taking the topic of Bigfoot have become more open-minded to discussing it.

Where did you grow up? In the Dallas Fort Worth area.

What are your hobbies? Cars, modifying them, driving them. Oh, and Bigfoot! Travel, especially CryptoTourism! Dining out, going to the movies.

What is your favorite musical genre/musicians/bands? The Cars! Rolling Stones, LED Zeppelin, INXS, Sarah McLachlan, Maria McKee, Lone Justice, AC/DC, Billy Idol

If you could take one person past or present on an expedition with you, who would it be? Adam Davies, he's a world-class explorer. Or Josh Gates for his sense of humor.

When and why did you get involved in Bigfoot research? I had a personal experience in 1994 and wanted answers.

Approximately how many eyewitnesses have you spoken to? 300+

What was the most compelling witness account that you heard first-hand? A minister that came so close to hitting a Bigfoot crossing the road in front of he and his wife that the Bigfoot's hand brushed against the hood of his truck. He said he regretted selling the truck as it had been touched by a Bigfoot.

What do you feel Bigfoot/Sasquatch are and why? An uncatalogued by science upright primate

What excites you about this subject? The discovery and cataloging as a species when it finally happens.

If Bigfoot is a large primate, what environmental conditions would it need to survive? Forested environment with 20+ inches of annual rainfall as theorized by John Green

How would you explain the elusiveness of Bigfoot? Stealthy, intelligent, not that many of them, maybe 2000-6000 in North America

What led you to put together the Texas Bigfoot Conference? I attended the Ohio Bigfoot Conference in 2000 and decided I could do this in Texas.

What were some memorable moments at the Texas Bigfoot Conference? Dr. Jeff Meldrum's reaction at an after-dinner event that was an Elvis concert performed by a Sasquatch enthusiast and professional Elvis impersonator. Meldrum said that he was just waiting for a UFO to land!

Also, we had a congressman, Louie Gohmert dropped by the conference. He was a guest of Dick Collins who sponsored Jeff Meldrum as a speaker. Jeff and I spent 20 minutes speaking with the congressman and his wife. He seemed genuinely interested in the topic of Bigfoot. Rep. Gohmert's work within the House Committee on Natural Resources has proved valuable to East Texas with its wide array of natural resources. We even discussed his going out in the field with us later that night with the news crew, unfortunately, he had other commitments.

The 2009 Texas Bigfoot Conference hosted probably the most prestigious speaker ever at a Bigfoot Conference, Peter Matthiessen, an American novelist, naturalist, wilderness writer, zen teacher and CIA officer.

What is the Texas Bigfoot Film Festival? An event where we screen multiple Bigfoot films in one day, with the directors and/or cryptozoologists on hand to introduce and discuss them. Last December we screened 5 films.

What impact has TV & social media had on the subject of Bigfoot over the recent years? It has been popularized with the general public and especially children.

What impact does hoaxing have and what do you feel motivates individuals to do so? Attention, and wanting to fit in.

Do you feel since the 1960s that we have gotten any closer to proving the existence of Bigfoot? We have collected evidence, but nothing as compelling as the P/G film

Why do you think there is not more evidence or proof? I feel they are elusive, intelligent, and rare animals.

How do you feel mainstream scientists will publicly accept the existence of Bigfoot/Sasquatch? A dead body to examine. Compelling video or photographic evidence that the source is unimpeachable.

What is your elevator pitch for open-minded skeptics? Look at the evidence that exists.

What suggestions do you have for researchers who want to learn how to get involved in this subject? Read, find a group of like-minded individuals.

How to follow Craig Woolheater:
Facebook: Texas Bigfoot Conference

CONCLUSION

It is evident that a phenomenon exists regarding the subject of Bigfoot. Thousands of reports have been documented including anecdotal evidence in the forms of visual, audio, and other types of witness accounts. In addition, hundreds of footprints, trackways, hair, and scat have been investigated which cannot be explained by any known animal in North America. Hairy, bipedal, unknown creatures are not exclusive to North America, reports have been documented worldwide. The witnesses come from various backgrounds and cultures. Some are trained observers like scientists, military, and law enforcement.

There are hundreds of individuals actively researching, creating media, or influencing enthusiasts on the subject. Their research takes many forms, and today's "citizen scientists" are focused and methodical. Many have dedicated years of their life and personal finances to try and solve the mystery. Sadly, some have passed away before getting closure.

I hope this book provided a glimpse into the current efforts being put forth to study the Bigfoot phenomenon and has given you additional resources to share your encounters or further your personal quest.

Will one of the individuals in this book find the proof of the elusive creature? You never know.

AFTERWORD

Go to hangar1publishing.com to learn more about the Author and stay up to date with their newest releases.

ADDITIONAL INFORMATION

Books Recommended by the Influencers

- Abominable Snowmen: Legend Come to Life - Ivan T. Sanderson
- Beyond Boggy Creek - Lyle Blackburn
- Bigfoot Evidence - Dr. Gover Krantz
- Bigfoot On The East Coast - Rick Berry
- Bigfoot the Life and Times of a Legend - Joshua Blu Buhs
- Descent of Man - Charles Darwin
- Giants, Cannibals and Monsters: Bigfoot in Native Culture - Kathy Strain
- Know the Sasquatch - Christopher L. Murphy
- Momo - Lyle Blackburn
- On The Track Of Bigfoot - Marion Place
- On the Track of the Sasquatch - John Green
- On the Track of Unknown Animals - Bernard Heuvelmans
- Origin of the Species - Charles Darwin
- Sasquatch - Rene' Dahinden
- Sasquatch, North Americas Great Ape - Dr. John Bindernagel
- Sasquatch: Legend Meets Science - Dr. Jeff Meldrum
- Sasquatch: The Apes Among Us - John Green
- Squatchin 101: How to Start Doing Your Own Research - Charles Kimbrough and Monongahela
- The Beast of Boggy Creek - Lyle Blackburn
- The Discovery of the Sasquatch - Dr. John Bindernagel

- The Hoopa Project - David Paulides
- The Locals - Thom Powell
- The Oregon Bigfoot Highway - Joe Beelart
- Tracking the Stone Man - Dr. Russ Jones
- Bigfoot! The True Story of Apes in America - Loren Coleman
- When Bigfoot Attacks - Michael Newton

Author and Publisher of the Bigfoot Times, Daniel Perez, has created *Project Bigfoot Books* and *The Center for Bigfoot Studies'* mission is to create the most complete listing of all known books on the subject of Bigfoot (aka Sasquatch), Yeti, Abominable Snowman, etc.

Project Bigfoot Books

Other Influencers

There are hundreds of individuals that continue to impact the quest for answers to the Bigfoot Phenomenon, here is a small compilation.

- Russ Accord
- Eric Altman
- Dave Bakara
- Paul Bartholomew
- Larry Batson
- Joe Beelart
- Chris Bennett
- Daniel Benoit

- Paul Bowman Jr.
- Seth Breedlove
- Reggie Byrd
- Dr. Angelo Capparella
- Shelly Covington-Montana
- Caroline Curtis
- Adam Davies
- Abe Del Rio
- Brown Family
- James Fay
- Henry Franzoni
- Ken Gerhard
- Wes Germer
- Bob Gimlin
- Paul Graves
- Mike Greene
- Todd Hale
- Jeff Harding
- Dr. Haskell Hart
- Jerry Hein
- Alton Higgins
- Ranae Holland
- Kenney Irish
- Don Keating
- Winona Kirk
- John Kirk
- Jen Kruse
- Charles Lamica
- Robert Leiterman
- Mike Mays
- Dave McCullough
- Bea Mills
- Matt Moneymaker
- Monongahela
- John Morley

- Marc Mursell
- Darren Naish
- Dan Nedrelo
- Christopher Noel
- Darby Orcutt
- Gareth Patterson
- Dave Paulides
- Jesus Payan
- Aleksander Petakov
- Thom Powell
- Derek Randles
- Charlie Raymond
- Robert Robinson
- Jim Sherman
- Pat Spain
- Chris Spencer
- Todd Standing
- Jeff Stewart
- Steven Streufart
- Lon Strickler
- Jim Thompson
- Josh Turner
- Tim Vogel
- Michael Waldie
- Ken Walker
- Doug Waller
- David Weatherly
- John Wilk
- Tom Yamarone

Bigfoot Research Organizations

American Primate Conservancy
Bigfoot Field Researchers Organization
East Coast Bigfoot Researchers Organization
Kentucky Bigfoot Researchers Organization
Minnesota Bigfoot Research Team
North American Bigfoot Center
North American Wood Ape Conservancy
Northern Kentucky Research Group
Squatchachusetts
Texas Bigfoot Research Organization
The Bluff Creek Project
The Olympic Project

Museums

Bigfoot Books
Bigfoot Crossroads of America Museum
Bigfoot Discovery Museum
Cryptozoology & Paranormal Museum
Expedition Bigfoot
International Cryptozoology Museum
Monster Mart
Museum of the Weird
North American Bigfoot Center
Sasquatch Outpost
Sasquatchthelegend.com
Skunk Ape Headquarters
West Virginia Bigfoot Museum
Willow Creek-China Flat Museum

Podcasts, Radio, Media

Apes Among Us (NAWAC)
Bigfoot and Beyond
Bigfoot Eyewitness Radio
Bigfoot Society
Bigfoot Terror in the Woods
From The Shadows
iNTO THE FRAY RADIO
Monster Talk
Monster X Radio
Monsters Among Us
Monstro Bizarro
My Bigfoot Sighting
Nite Callers Radio
Phantoms & Monsters
Sasquatch Chronicles
Sasquatch Odyssey
Sasquatch Out of the Shadows
Sasquatch Tracks Podcast
Squatch-D TV
The Bigfoot Influencers (coming soon)
The Bluff Creek Project
The Confessionals
The Cryptid Factor
The Forest Fluer
Untold Radio AM
What If It's True
Wild Thing

Here is a Small List of Television Series, Documentaries, and Movies

Bigfoot: Man or Beast (1972)
Exists
Expedition Bigfoot
Finding Bigfoot
Harry and the Hendersons
Letters from the Big Man
MonsterQuest
Mountain Monsters
Sasquatch Odyssey: The Hunt for Bigfoot
Sasquatch: Legend Meets Science
Small Town Monsters
Survivorman Bigfoot
The Dark Divide
The Legend of Boggy Creek
Willow Creek

ACKNOWLEDGMENTS

Writing this book has been truly a team effort. Starting with the creative support, introductions, and editing, much appreciation to Matt Pruitt, Doug Hajicek, Les Stroud, Alex Hajicek, Blaine Hajicek, Daniel Perez, Russ Jones, and Hangar 1 Publishing for your creative input and support for the book.

My team of volunteer editors and proofreaders; Amy Bue, Dana Halloran, John Adams, and Danny Halloran. Gabriel DaSilva for helping to create the interview questions and Tony Halloran for listening to my recap and theories from my conducted interviews. Noah Cicogna for sharing an office with me and enduring through my conversations related to the book. The rest of the family for your open-mindedness and support; Hugh, Lee, Carol, Rich, Jen, Rob, Donna, Wayne, Dave, and Abby.

Thank you, Todd Neiss and Daniel Perez, for the introduction to Peter Byrne. Marc DeWerth, for allowing me to take part in the Ohio Bigfoot Conference, granting me the opportunity to meet many impactful researchers in this field.

For the scientists and researchers that participated in this book, I am grateful for the opportunity, your insight and diligence in the pursuit of answers is inspiring. This has been truly a team endeavor and you are equally part of this book project as am I. It has been a joy getting to know each of you and a tremendous learning experience for me. I hope this book provides positive insight into your research and efforts.

ABOUT THE AUTHOR

This is a debut book for author Tim Halloran, who grew up in Maryland, just outside of Washington DC. He and his wife Dana currently reside near the coastal region of Delaware with two of their four boys, two cats and whatever wildlife may show up in the yard. Away from his career and writing this book, Tim spends as much of his spare time as possible outside, he and Dana enjoy, hiking, gardening, working on home projects, and live music.

The Bigfoot phenomenon was not on Tim's radar prior to Dana convincing him to go to conferences. Though not a researcher himself, and not planning on writing a book on this subject, the idea was inspired by the fascination of the dedicated groups and individuals who have the common quest of solving the Bigfoot mystery.

http://thebigfootinfluencers.com
Facebook: @thebigfootinfluencers
Instagram: @thebigfootinfluencers
Twitter: @BFInfluencers
Email: thebigfootinfluencers@gmail.com

Social Media Links

www.ingramcontent.com/pod-product-compliance
Lightning Source LLC
Chambersburg PA
CBHW070057030426
42335CB00016B/1922
* 9 7 8 1 9 5 5 4 7 1 5 8 9 *